A
MOST
DAMNABLE
INVENTION

———

ALSO BY STEPHEN R. BOWN

*Scurvy: How a Surgeon, a Mariner, and
a Gentleman Solved the Greatest Medical Mystery of
the Age of Sail*

*The Naturalists: Scientific Travelers in
the Golden Age of Natural History*

A
MOST
DAMNABLE
INVENTION

*Dynamite, Nitrates, and
the Making of the Modern World*

STEPHEN R. BOWN

THOMAS DUNNE BOOKS
St. Martin's Press New York

THOMAS DUNNE BOOKS.
An imprint of St. Martin's Press.

www.stmartins.com
Design by Level C

Library of Congress Cataloging-in-Publication Data

Bown, Stephen R.
    A most damnable invention : dynamite, nitrates, and the making of the modern world / Stephen R. Bown.—1st ed.
        p.   cm.
    Includes bibliographical references (p. 251) and index (p. 259).
    ISBN-13: 978-0-312-32913-6
    ISBN-10: 0-312-32913-X
    1. Science—Moral and ethical aspects.   2. Science—History.   I. Title.
Q175.35.B69 2005
174'.95—dc22

2005045527

First Edition: October 2005
10 9 8 7 6 5 4 3 2 1

Contents

Acknowledgments

Many people have contributed to this book from its conception through to its publication. I would like to thank Peter Wolverton for his thoughtful editorial suggestions on my first draft, and Donald J. Davidson for his thorough and informed copyedit. Thanks also to Robert Sessions, Saskia Adams, Diane Turbide, Bob Berkel, and Katie Gilligan. My agents, Frances and Bill Hanna, were invaluable in many ways, from convincing several publishers around the world to take on this project, to editorial commentary, to helping to keep the whole thing coordinated.

Once again I'd like to acknowledge the help I've received from the staff at the Canmore Public Library in acquiring odd, obscure, and difficult-to-obtain books for me; also the Alberta Foundation for the Arts and the Canada Council for the Arts. My brother David Bown designed my author Web site and provided many hours of help with computer-related issues. Thanks also to my brother Mike Bown for putting me on to part of this story when he mentioned an interesting article he had read about the use of Peruvian guano for explosives. Most important, as always, I am grateful to my wife, Nicky Brink, for hours of conversation while I sorted my thoughts, perceptive reading of my first draft, general encouragement, and for keeping our two young children entertained when I had a deadline.

Time Line

424 B.C.	Greek fire is used at the Battle of Delium during the Peloponnesian War.
718	Greek fire is used to defend Constantinople from an Islamic invasion fleet.
941	Greek fire is used to defend Constantinople from an invasion by Prince Igor of the Kingdom of Kiev.
1100s	Firecrackers are used in China.
1200s	Buddhist cave engravings in Szechuan Province in China show the use of gunpowder weapons.
ca. 1249	The English monk Roger Bacon records the secret of gunpowder.
ca. 1300s	Markus the Greek writes *Book of Fires for Consuming the Enemy*.
1346	The Battle of Crécy in France is fought wherein English longbowmen triumph over mounted knights and crossbowmen.
1449	Long-range mobile cannons prove superior to the English longbow and the English are expelled from France.

1453	Constantinople is conquered by Mohammed II and his massive siege guns. The Hundred Years' War end. The Middle Ages end.
1492	Christopher Columbus first crosses the Atlantic from Spain to the Caribbean, beginning the great era of European naval exploration.
1519–1532	The Spanish conquest of the Aztecs and Incas is greatly aided by guns.
1526	Moghul armies under Babar invade northern India and conquer it using cannon.
1588	The Spanish Armada fails in its attempt to invade England.
1600	Tokugawa Ieyasu begins the restriction of gunpowder weapons in Japan.
1605	Guy Fawkes attempts to explode the English Parliament with kegs of gunpowder.
1626	Charles I of England proclaims that all citizens must collect their urine for saltpeter production.
ca. 1650s to early nineteenth century	Indian saltpeter from Bihar and Bengal fuels European wars.
1681	Gunpowder is used to construct the Languedoc Canal in France.
1700s	Dutch traders dominate the Indian saltpeter trade.
1734	Carolus Linnaeus records the foul treatment of workers at the Falu copper mine in Sweden.

1756	Beginning of the Seven Years War between European powers, with England and France on opposing sides. Samuel Johnson's *Dictionary* is published.
1757	At the Battle of Plassey, troops of the English East India Company defeat the French in India. English colonial rule in India begins.
1760s	Britain begins consolidating control over India and the saltpeter regions of Bihar and Bengal and regulating saltpeter shipments to other nations.
1768–1771	Lieutenant James Cook leads his first voyage of discovery.
1775–1783	The War of American Independence is fought.
1779	The great sea captain James Cook is killed in Hawaii.
1782	James Watt invents the double-acting rotary steam engine.
1805	Nelson wins the Battle of Trafalgar, destroying Napoleon's fleet.
1815	Napoleon is defeated at the Battle of Waterloo.
1817–1825	Gunpowder is used to remove obstacles to the Erie Canal, which linked the Hudson River with the Great Lakes, making westward expansion possible.
1831–1836	The *Beagle* departs England for South America and the Galapagos with Charles Darwin as naturalist.
1830s	Peruvian guano rapidly becomes indispensable as a fertilizer.
1834	Alfred Nobel, the inventor of dynamite, is born.

1846	Ascanio Sobrero invents nitroglycerin. Christian Friedrich Schönbein invents guncotton.
1853–56	Britain, France, and the Ottoman Empire halt Russian southward expansion during the Crimean War. Alfred Nobel works for his father's business designing and manufacturing armaments for the Russian government.
1856	The Guano Island Annexation Act is enacted by the United States.
1857	The discovery of potassium deposits in Germany makes the conversion of Chilean caliche saltpeter into gunpowder easier and cheaper.
1861–1865	The War Between the States is fought in America.
1867	Nobel invents dynamite. The Suez Canal is completed.
1868	Fritz Haber is born.
1870s	Shipments of Chilean caliche saltpeter rise dramatically after Nobel's invention of dynamite.
1870–1871	The French are defeated in the Franco-Prussian War, in part, because they lacked dynamite.
1871	The British Dynamite Company is founded in Scotland
1873	Nobel moves to Paris.
1875	Nobel invents blasting gelatin (gelignite).
1876	Hallett's Point reef is blasted in New York harbor. The Hoosac Tunnel is completed.
1879–1884	Bolivia, Peru, and Chile fight the War of the Pacific over control of the caliche nitrates in the Atacama Desert.

1880	The St. Gotthard Tunnel is completed, linking Italy and Switzerland.
1883	Australia consumes 50 percent of the dynamite used in the British Empire.
1886	The Severn Tunnel, the longest railroad tunnel in Great Britain, is completed, linking England and Wales. Paul M. Vielle invents Poudre B in France. Nobel invents Ballistite, a form of smokeless powder, which quickly replaces black powder.
1880s–1890	Frederick Abel and James Dewar develop cordite, another smokeless powder, in England.
1890s	Chilean nitrates supply two-thirds of global demand.
1891	Nobel relocates his home and lab from Paris to San Remo, Italy.
1893	The Corinth Canal completed, linking the Ionian and Aegean Seas.
1896	Alfred Nobel dies in San Remo, Italy, and leaves his entire fortune to fund international prizes.
1901	The first series of Nobel prizes is awarded.
1904	The first branch of the New York City subway opens.
1906	The Simplon Railway Tunnel, one of the longest in the world, linking Italy and Switzerland is completed.
1909	Haber's experiment produces a synthetic nitrogen compound for the first time.
1913	Carl Bosch rapidly adapts Haber's experimental model for use in a viable industrial plant at Oppau.

1914   The Panama Canal is completed. The Battle of Coronel between German and British fleets takes place off the coast of Chile. The German victory cuts Britain off from its nitrate supply. The Battle of the Falklands establishes British naval superiority and begins the naval blockade of Germany.

1914–1918   Bosch rapidly expands the plant that supplies Germany with the bulk of its nitrate needs for ammunition and fertilizer throughout the war.

1915   Haber designs and organizes the first chlorine gas attack at Ypres.

1918   The First World War ends.

1919   Fritz Haber wins the Nobel Prize in Chemistry.

1934   Haber dies an exile in Switzerland.

1919–2005   Global population triples after the invention of synthetic nitrogen fertilizer, which exponentially increases food supplies.

A
MOST
DAMNABLE
INVENTION

———

An Epic Quest

Fritz Haber was in Stockholm to receive a lucrative and prestigious prize. It was June 1, 1920, and he was fifty-two years old. The award of a Nobel Prize in Chemistry ought to have been cause for celebration, a crowning recognition of his brilliant scientific work. But a shadow hung over the ceremony. The king of Sweden did not present the award, as he had done for the four other awards the previous November. Haber received his prize alone half a year later, the only exception since the inception of the awards fifteen years earlier, because of public outrage. French, American, and British scientists denounced Haber's selection as prizewinner, arguing he should have been ineligible for distinction due to his dishonorable, perhaps even immoral, wartime activities. They labeled Haber "the inventor of gas war."

None disputed his scientific credentials, however. Haber's work was universally recognized as being of profound importance, perhaps even the solution to world hunger. He had solved the nitrogen problem, the Holy Grail of scientific enigmas of the era, patenting a process whereby nitrogen compounds could be synthesized from the air, and had thereby set the stage for the interminable war of attrition between 1914 and 1918. Two German chemical plants, at Oppau and Leuna, based on Haber's design, a triumph of scientific and technical genius for their time, were the secret weapons in Germany's arsenal that kept the nation fighting during the last years of the First World War by supplying at once the raw material for explosives and fertilizers. Haber's scientific breakthrough was also the conclusion of a centuries-long quest to sever the tether that bound farmers, miners, and warriors to a scarce and somewhat elusive organic source of the mundane substance that is vital in improving crop yields in agriculture, yet is also at the heart of all explosives.

Humanity's quest to harness the destructive capacity of fire is a saga that extends back to the dawn of civilization. The use of highly combustible black powders—a crude blend of sulfur, charcoal, and saltpeter—spread west from China through the Middle East, arriving in Europe in the late thirteenth century. Influence in world affairs, the protection of trade routes, and resistance to being conquered soon depended upon having full use of supplies of black powder, which began to be called gunpowder. Although gunpowder did bring about social change, toppling feudalism and ushering in a new military structure, there was always a shortage when it was most needed, and it was never powerful enough to fully enable the dreams of ambitious men. The true great era of explosives, when they radically and irrevocably changed the world, began in the 1860s with the remarkable intuition of a sallow Swedish chemist named Alfred Nobel.

Nobel made a discovery that refocused the direction of explosives research, altering centuries of work on refining black powders and opening a new frontier in the development of vastly more powerful "high" explosives. He sought and received a patent for his discovery and soon established factories across Europe and the United States. He struggled for years with industrial accidents that killed dozens of his employees and customers before he finally settled on a method to secure the safety of his product for storage and transportation. His new and improved product would have an immediate and profound impact upon western Europe and the United States and eventually throughout the world.

Within a decade, he was one of the richest men in a society rapidly transforming under the power of his invention. Nobel's breakthrough discovery was how to stabilize and harness the explosive power of nitroglycerin, a mixture of sulfuric acid, nitric acid, and glycerin. He named the malleable putty dynamite, after the Greek *dynamos*, meaning "powerful," and perfected a reliable method of detonating it. Dynamite was the first truly safe and predictable explosive that had a strength vastly superior to gunpowder. It was one of the most significant technical developments in a century known for revolutionary inventions.

Dynamite swiftly transformed industry and warfare, becoming indispensable to both. It unleashed the power that enabled the monumental industrial development of the late nineteenth and early twentieth centuries, including hydro power, skyscrapers, and coal and oil exploration. Canals and railroads, mining and construction, tunnels and harbors, land mines and artillery: The explosive power of dynamite contributed to the burgeoning Industrial Revolution. Even concrete could become a common construction material once limestone could be blasted from quarries. Dynamite liberated laborers from the tiresome and unremitting toil of slaving in mines and quarries or moving and leveling earth. What once took weeks, or was possible only in the fanciful dreams of the overly optimistic, could now be done in moments. Dynamite and its derivatives were also used to construct greater and more powerful military tools, from land mines and bombs to artillery shells.

In the second half of the nineteenth century, however, science and industry remained tethered to the practical reality that no explosive could be made without the organic compound known colloquially as saltpeter—the crystals that form from decomposing vegetable and animal matter in hot, dry earth. Saltpeter (usually potassium nitrate, but also sodium nitrate or ammonium nitrate) was the ingredient most scarce and difficult to obtain, and as such it was sought after and coveted because there was never enough, particularly during times of war. The enduring quest for a stable and secure supply of nitrates has had almost as great an impact on world affairs as the explosives themselves. It is a story that takes us from the rural stables and privies of preindustrial Europe to the monopoly trading companies and plantations in eighteenth-century India to the guano islands off Peru to the Atacama Desert in northern Chile and the War of the Pacific in 1880. Nitrates were a substance as valuable in the seventeenth, eighteenth, and nineteenth centuries as oil is in the twenty-first, and the cause of similar international jockeying and power politics. By the early twentieth century, the only commercially viable organic source of nitrates in the world that could meet the escalated demand brought about by the remarkable proliferation of high explosives and the dramatic increase in its use for agriculture was in Chile, halfway around the world from its primary markets in western Europe and the eastern United States.

Explosives had become so vital to industry and the military, and fertilizers so critical to the agricultural economy, that maintaining access to the Chilean nitrate deposits was of strategic importance, and a legitimate concern for national defense. When war broke out in 1914, securing access to the Chilean nitrates, and denying that access to others, was one of the most pressing objectives in the early months of the conflict. A British naval victory seemed to presage an early end to the war by closing the sea-lanes to Germany and strangling at once both its food production and its production of armaments. But Haber's timely and brilliant scientific discovery on the eve of the war changed the course of the war and of world history.

Haber's work also provided the foundation for the explosive capacity of the bloodiest century in our history. There is a certain irony in Haber's being awarded the Nobel Prize for work which, although continuing Nobel's own work on explosives (the foundation of the prize money in the first place), has unleashed an unlimited quantity of explosives for war and ultimately caused the deaths of millions of people—something that would have appalled Nobel had he lived to witness it. Although it was war that inspired Haber's chemical discovery, its greatest benefit has been in agriculture, in the creating of synthetic fertilizers, one of the most significant, if underappreciated, advance of the twentieth century.

Next to Nobel's invention of dynamite, it is hard to imagine another single technological discovery or innovation that has had as long, and as lasting, an impact on human affairs and the shaping of our physical and social environment. Without Nobel's dynamite our modern economy would not exist, while Haber's creative genius has had unfathomable repercussions for agriculture and global population in the century that followed. The story of Nobel and Haber and their scientific innovation is one of the epic stories of human accomplishment. It is the tale of a remarkable technology, the historic impact of that technology, and the globe-spanning struggle for the raw material required to create and make use of it.

Playing with Fire

A Thousand Years of Explosives

It hath been doubted whether so ingenious and dreadful a Ma-chine could be a humane Invention . . . when it was first pub-lished, the World thought she had lost all her strength; for what more terrible or violent could humane Wit invent to its own de-struction, than this artificial Lightning and Thunder.
—William Clarke, 1670

Four centuries ago, on January 31, 1606, a bedraggled, haggard, and limping man was led through the gathered throng at the Old Palace Yard at Westminster. Past the laid-out corpses of his com-rades and fellows, past the solemn deputies and mounted justices and sheriffs, past the pike-bearing men-at-arms who kept the surging onlookers at bay. He was roughly hauled up the stairs onto a newly constructed scaffolding in the center of the square and turned to face the black-hooded hangman. Crowds congregated in the vast courtyard, hoping for a good view of the action; ven-dors sold foods and beer. Whether the day was pervaded by an air of lighthearted gaiety or profound import is not reliably docu-mented, but it was a significant moment in England's history, and the people knew it. The man's imminent death was a cause for cel-ebration, however muted during the final moments. After making a short speech, the hunched, red-bearded man bowed weakly for the noose, slowly crossed himself, and prepared to die.

The man's name was Guy Fawkes and his crime was none other

than high treason. A quick and foreordained trial had found him and a handful of others guilty of a most horrendous and frightening crime: the attempted assassination of the new king, James I, his queen, and the lords of the realm while they gathered in the House of Lords for the opening of Parliament. The new king had only ascended to the throne of England after the death of Elizabeth a few years before. Fawkes had signed his confession, his own death warrant, in a crippled, barely legible scrawl that reveals the extent of his torture. After nearly three days in the Tower of London, stretched on the rack and mercilessly squeezed by the manacles to extract a suitable statement, he had broken down and revealed the details of his bold and fiendish plot.

Nothing in Fawkes's upbringing destined him for notoriety and revulsion as a national traitor and potential murderer. The date of his death is still commemorated four centuries later. Born into a respectable family in York in 1570, he enjoyed a good education and a comfortable upbringing. His father died when he was eight, and for nine years he was raised by his mother. She remarried when he was seventeen, and her new husband was a recusant gentleman who first introduced Fawkes to the Catholic faith. Fawkes came of age amid the religious turmoil of the Protestant Reformation, a northern European religious movement that sought to shake off the bonds of papal authority. England was the latest in a series of countries, centering on Switzerland, Scandinavia, Scotland, and many German states, that were frustrated by the corruption of the church. Henry VIII established the Anglican church and began confiscating Roman Catholic church property and dissolving monasteries. For much of the second half of the sixteenth century opposing factions struggled to place either Catholic or Protestant monarchs on the English throne. When Elizabeth I ascended to the throne in 1558, conflict with Catholic Spain escalated and the plight of Catholics in England grew worse. Many of the Catholic families had their lands confiscated, and they were driven underground, to practice their religion in secret to avoid reprisals. When Philip II of Spain launched his grand armada in 1588, it was an attempt to depose Elizabeth and set a Catholic monarch on the throne and outlaw the Protestant faith. It was a savage and barbarous age. The separation of church and state and religious tolerance, even between Christian factions, lay in the distant future.

At some point the young Fawkes converted to Catholicism under the influence of his stepfather and decided to leave England for Flanders, where he enlisted in the Spanish army (then occupying the Netherlands in a bloody war to crush the growth of Protestantism). He was reputedly a strong and calm commander "of excellent parts, very resolute and universally learned." He earned a reputation as "a man of great piety . . . remarkable for his punctual attendance upon religious observance." After he had served with distinction in the Spanish army for years, the issue of religion in England again raised its ugly head. Elizabeth I died in 1603, and the throne passed to the Scottish king James V. English Catholics in exile agreed to send Fawkes on a mission to Spain to obtain support for another Spanish invasion of England, claiming that the English people would eagerly rise up and overthrow the new king, now James I. When his plea failed, Fawkes returned to Flanders and met with several other fanatical men who resolved on a plan to murder the new king themselves in defense of their religion.

The ringleader of the small group was a recusant country gentleman from Warwickshire named Robert Catesby. He urged blowing up Westminster because "in that place, they have done us all the mischief and perchance God hath designed that place for their punishment." At first the plotters began digging a tunnel, and when this became too difficult, they rented a vacant storeroom under the Parliament that at one time had been used to store coal. They then somehow secretly ferried thirty-six hundred-pound barrels of gunpowder across the Thames from Catesby's house and trundled it down into the storeroom and covered the mountain of casks with firewood. The opening of Parliament was originally scheduled for February 1605 and was then postponed until October 3 and then again to November 5. Fawkes, who had taken on the role of firing the powder because of his military experience, occasionally checked it and replaced any that had become too "decayed" to explode after such long storage in the damp cellar.

Meanwhile, Catesby had widened the plot and taken in more conspirators. They planned to flee to Europe soon after the explosion to spread the good news and rally support for an uprising among Catholic troops stationed in Flanders. On Saturday October 26, an anonymous servant delivered a nondescript letter to William Parker, Lord Monteagle, as he settled down for dinner in the evening (Monteagle was married to the sister of one of the newly recruited conspirators). "This Parliament," the letter read, "shall receive a terrible blow, and yet they shall not see who hurts them." Monteagle immediately raised the alarm. A thorough, yet discreet, scouring of the premises revealed the unusually large pile of firewood in the cellar and a man claiming to be John Johnson guarding it. Johnson, who was actually Fawkes, was apparently "a man shrewd enough, but up to no good." He was immediately arrested and dragged away for questioning. He later claimed that had he been prepared or quick enough when the guards entered the cellar, he would have "blown him up, house, himself, and all."

Later that night Fawkes was presented to the king and asked why he took part in such a cowardly scheme. "A dangerous disease required a desperate remedy" was his defiant reply. The next morning James I issued a letter instructing his constables that "The gentler tortours are to be first used unto him, and so by degrees proceeding to the worst, and so God speed your goode worke." After three dreadful days a mangled and broken Fawkes confessed all, revealing the names of the other conspirators and that his motivation was "for the advancement of the Catholic Faith and saving his own soul." After a swift mock trial he was led to his death. When he was jerked into the air and swung from the gibbet, it was, according to the official government propaganda of the time, "to the great joy of all the beholders, that the land was ended of so wicked a villainy." In quick succession several others were similarly hanged, their bodies left to swing erratically, before the ropes were cut and they dropped unceremoniously to the earth. A contemporary engraving of the scene shows the curious throngs ringing the courtyard while pike-wielding soldiers keep order. The still live bodies of the conspirators were dragged by horses around the clearing (drawing) before being castrated, eviserated, and beheaded. Each limp limb was tied to a stout horse, which then surged forward at a gallop, tearing the corpse into four parts (quartering), according to the brutal custom of the day. The date of Fawkes's capture, November 5, 1605, was proclaimed a public holiday, which is still celebrated today with firecrackers and the burning of effigies. One of the liturgies of the Church of England soon afterward was titled *A form of prayer with thanksgiving to be used yearly upon the fifth day of November; for the happy deliverance of the King, and the three estates of the realm, from the most traiterous and bloudy intended massacre by gun-powder.*

Historians have since questioned the accuracy of the official government story of the time. The scene of the fiendish, mustachioed Fawkes hunched over a burning slow match adjacent to a great mound of gunpowder, cackling and singing his defiance of the king and the Protestant faith, seems a little too contrived. Many historians now believe that the plot was as much a government scheme to flush out disloyal Catholics as a secret plot to blow up Parliament, or alternately that the brash public display of the execution was orchestrated to conceal how easily the plot almost succeeded and how easily the government monopoly on gunpowder production was circumvented.

Up until this time the potential of gunpowder was still being explored and was not entirely appreciated. Although it had been used in guns and cannons for several centuries, their effectiveness was only slowly improving to the point where the damage they inflicted was equal to the frightening noise and billowing smoke. Fawkes's creative and new use of gunpowder as a targeted explosive outside of cannons foreshadowed the tremendous power that would in the late nineteenth century be easily available in the form of dynamite and other high explosives. The Gunpowder Plot for the first time revealed just how powerful black powder could be and starkly exposed its awesome capacity to play kingmaker and determine the destiny of nations. In the political turmoil and uncertainty following Queen Elizabeth's death, and the ongoing religious struggles between Catholics and Protestants, the near miss of the plot, how close the plotters had come to eliminating a government they despised, turned heads across Europe. If the relative simplicity of the plan became generally known, future plots by religious zealots and usurpers would be assured. The origin of the vast quantity of gunpowder placed by Fawkes under the Houses of Parliament was never mentioned in the official records of the trial, probably to conceal just how easily such a frightening and remarkably dangerous substance could be obtained by amateurs, even though it was technically under government control. It was one thing for an assassin to attack by wielding a knife or by shooting a crossbow or placing poison in food, quite another for a handful of disgruntled fanatics to bring down the entire government with ease and without a moment's notice.

———

This contemporary sketch shows Guy Fawkes and his fellow conspirators in the Gunpowder Plot to explode the English parliament in 1605.

Three and a half centuries before Guy Fawkes tried to explode the English Parliament a curmudgeonly middle-aged English scientist and friar named Roger Bacon staggered back from a terrifying explosion in his Oxford monastery cell. A jarring boom was followed by a billowing cloud of noxious gases from a crucible on the table where he had been experimenting. Coughing and choking on the brimstone fumes and shaken by the lightninglike eruption, he quickly moved to conceal the evidence of his experiment. His haste and secrecy were prompted by fear both for the destructive capacity he had unleashed and for his own safety if he were branded a heretic for toying with magic or matters of the devil. Unlike his other discoveries, which he wrote about in great detail, his experiments with black powder he concealed and only alluded to in his writings. Bacon was one of the most famous and forward-thinking philosophers of his time, and one of the great practical experimenters of the medieval era.

Born at Ilchester in Somerset into a wealthy family in 1214, Bacon quickly earned a reputation as a bold and original thinker. After receiving his degrees from Oxford, he studied and taught at the University of Paris, then the center of learning in the European world, where he earned the title "Doctor Mirabilis" (Astounding Doctor) for his brilliant theorizing and knack for getting to the heart of matters. After a distinguished career at the university, Bacon joined the Franciscans in the 1250s, probably for health reasons, and returned to England. His passionate abhorrence of dogma and superstition soon ran him afoul of the church authorities—a conflict the deeply religious friar struggled with his entire life. Bacon was a participant in the quarrels between science and religion that would dominate much of European thought for centuries thereafter. Some historians consider him to be the first modern scientist. He earned a reputation as a sorcerer, alchemist, and magician despite his lifelong insistence on empiricism and objectivity.

A deeply religious and devout Christian, Bacon passionately believed that a failure to explore the world was an insult to God, and that it was humanity's obligation to study nature. "If someone who has never seen fire," he wrote, "claims through reasoning that fire burns, changes things and destroys them, the mind of his listener will not be satisfied with that, and will not avoid fire before he has placed his hand or something combustible on the fire, to prove through experience what his reasoning had taught him. But once it has had the experience of combustion the mind is assured and rests in the light of truth. Reasoning is not enough—one needs experience." In addition to suggesting a much needed calendar reform to the church, he set out to show that seemingly supernatural phenomena were not the product of dark magical forces, but were rather products of the natural world and reproducible by experimentation. Two centuries before Leonardo da

Vinci, he predicted the development of "things of Nature that will amaze and astonish us" such as "perpetuall lights, and baths burning without end," spectacles, telescopes, magnifying glasses, flying machines, and motorized ships. In 1266, many of these observations were published in his *Opus maius,* a tome only recognized centuries later for its astonishing modernity. "These are marvailous things," he wrote, "if men knewe how to use them effectually in due quantitie and matter."

Bacon's most famous discovery was the recipe by which "the sound of thunder may be artificially reproduced by natural causes." In the Middle Ages, despite Bacon's insistence otherwise, the study of science was often blended with a belief in magic powers. Although the church was naturally suspicious of this line of study, Bacon was a keen alchemist and believer in the philosopher's stone, the universal secret to turning ordinary metals into gold. Most of his experiments involved a primitive form of trial-and-error chemistry, which is probably how he stumbled upon the formula for black powder. Bacon had been scouring ancient Arabic texts and had come upon a recipe for exploding powder, which was at one time believed to be related to the elixir of immortality. Bacon instinctively grasped that the practical application of this black powder would not be for the furtherance of science but as a tool in war. "Noyses may bee made in the aire like thunders," he wrote in *The Mirror of Alchemy.* "Yea with greater horror then those that come by Nature: for a little matter fitted to the quantitie of a thumbe, maketh a horrible noyse, and wonderfull lightning . . . whereby any citie and armie may be destroyed."

Out of fear of persecution and the violent potential of his discovery, Bacon concealed his recipe for black powder in a cryptic Latin code that was not deciphered for over six centuries. Bacon continued to pursue his alchemical studies, repeatedly circumventing attempts to censure his experiments and writings. Boldness and independence were not traits appreciated in the thirteenth century, and in 1277 he was imprisoned by Jerome of Ascali, minister general of the Franciscans. "By the advice of many friars," reads the pronouncement, "condemned and denounced the teaching of Roger Bacon of England, master of sacred theology, as containing some suspected novelties, on account of which the said Roger was condemned to prison, with the order to all brethren that none should hold his doctrine but avoid it as reprobated by the Order." He was released fifteen years later, a year before his death in 1292 at the age of seventy-eight.

Within a half century of Bacon's death, gunpowder was well on the road to common use in the military; its civilian applications in mining and road building were still centuries in the future. Although Bacon is given credit for the first written description of black powder in Europe, it had been in common use for centuries in Asia.

The quest to harness the destructive power of fire and explosives is as old as civilization. The earliest known incendiary is Greek fire, a term used to describe many different substances that blended sulfur with tar, rosin, bitumen, or other combustibles. Greek fire was first used in the Aegean region in the fifth century B.C. in naval battles and to burn city gates. Thucydides describes the use of flaming blow tubes in the battle of Delium in 424 B.C. during the Peloponnesian war between Athens and Sparta. "They took a great beam, sawed in two parts, both of which they completely hollowed out, and then fitted the two parts closely together again, as in the joints of a pipe. A cauldron was then attached with chains to one end of the beam, and an iron tube, curving down into the cauldron, was inserted. . . . When this machine was brought up close to the city wall, they inserted into their end of the beam large bellows and blew through them. The blast, confined inside the tube, went strait into the cauldron which was filled with lighted coals, sulfur and pitch. A great flame was produced which set fire to the wall and made it impossible for the defenders to stay at their posts. They abandoned their positions and fled; and so the fortification was captured." The use of Greek fire and similar substances remained the most terrible weapon for centuries and was adapted for naval use when kegs of sulfur, pitch, and tow were catapulted toward enemy ships where they exploded into flames that were exceptionally difficult to extinguish as they spread on the water. During the seventh century A.D. a Greek or Syrian alchemist and architect named Kallinikos, or Callinicus, improved upon the recipe and potency of the original "wild fire" by blending sulfur, crude naphtha, and quicklime—a toxic brew that formed an explosive gas that spread rapidly when exposed to water and then air.

Greek fire proved to be so powerful, effective, and frightening that its manufacture was a tightly held secret that kept the Islamic invaders from plundering the legendary city of Constantinople, the capital of the Byzantine Empire, for centuries. In 718 a grand Islamic invasion force under the caliphs of Damascus, with a fleet numbering over a thousand galleys, was besieging Constantinople and was virtually destroyed by the terrible oily flames that spread across the water and burned the hulls of the ships. The Roman historian Pliny recorded the use of Greek fire in *The Historie of the Worlde*, "Now let us relate some strange wonders of fire also, which is the fourth element of nature. But first, out of waters. In a citie of Comagene, named Samosatis, there is a pond, yeelding forth a kind of slimie mud (called Matha) which will burne cleare. When it meeteth with any thing solide and hard, it sticketh to it like glew: also if it bee touched, it followeth them that flee from it. By this meanes the townesmen defended their walls, when Lucullus gave the assault, and his souldiours fried and burned in their owne armours." Greek fire also saved Constantinople from an invasion from the north in 941 when Prince Igor of the Kingdom of Kiev launched a fleet across the Black Sea to raid the great city. According to the Russian *Primary Chronicle,* "The Greeks met them in their boats and began to shoot fire through pipes onto the Russian boats. And a fearsome wonder was to be seen. . . . The Greeks possess something like the lightning in the heavens, and they released it and burned us." During the Crusades, Islamic defenders hurled small glass or clay pots filled with flammable petroleum and lime mixtures down on Christian invaders.

In the thirteenth century, around the time of Roger Bacon's death, an unknown quasi-mythical figure named Markus the Greek published a book on pyrotechnics called *The Book of Fires for Consuming the Enemy.* The thirty-five-page tract is a list of recipes for various types of Greek fire that probably had been used for centuries before he recorded them. Incendiary weapons took many surprising and terrifying forms by the time of Markus the Greek. One of the more outlandish recipes contained in *The Book of Fires* describes how burning birds can be effective weapons. "Another kind of fire for burning enemies wherever they are," reported the author, "can be made by taking petroleum, liquid pitch, and oil of sulphur. Put all these in a pottery jar buried in horse manure for fifteen days. Take it out and smear with it crows which can be flown against the tents of the enemy. When the sun rises and before the heat has melted it the mixture will inflame. But we advise that it should be used before sunrise or after sunset." Presumably the burning birds would ignite the tents where the enemy soldiers slept and consume them in a fiery mass of burning fabric, or at least cause confusion and disturb their slumber. Fire birds are also mentioned in early Chinese and Arabic writings. The world's first mobile battlefield flamethrower was perfected in China by the early fifteenth century. Dozens of arm-length bamboo tubes, filled with a flammable concoction, were suspended by a wheeled three-meter-high rack that was rolled in front of marching troops. The tubes were ignited in sequence, with the objective of decapitating "the enemy soldiers, and to cut off the legs of their horses," according to a military treatise called the *Fire-Drake Manual*. "A single one of these shields is in itself worth ten brave soldiers."

The first primitive black powders also had their origin in China, in the form of firecrackers and smoke signals, as early as the tenth century. The historical record is sketchy and incomplete, and historians disagree whether the Chinese were merely toying with a combustible smoke-producing powder or had actually discovered the explosive black powder. If they had come upon such a deadly military weapon it would have remained a secret as long as possible to preserve their technological advantage. By the mid-eleventh century, however, the production of sulfur and saltpeter was under state control and their sale to foreigners was forbidden. Written descriptions of celebrations from the thirteenth century describe flaming balls and sky-borne sparks and flashes accompanied by thunderous noise. One account from 1264 reveals how uncontrollable the early fireworks were: "When the Emperor Li Tsung retired, he prepared a feast in honor of his mother, the Empress-Mother Kung Sheng. A display of fireworks was given in the courtyard. One of these, of the 'ground rat' type, went strait up the steps to the throne of the Empress-Mother, and gave her quite a fright. She stood up in anger, gathered her skirts around her, and stopped the feast. Li, being very worried, arrested the officials who had been responsible for making the arrangements for the occasion, and awaited orders from the Empress-Mother. At dawn the next day he went to apologize to her, saying that the responsible officials had been careless, and took the blame on himself. But the Empress-Mother laughed and said, 'That thing seemed to come specially to frighten me, but probably it was an unintentional mistake, and it can be forgiven'."

Mongol armies invading Poland and Hungary in the late thirteenth century reportedly used primitive bombs consisting of bamboo tubes loaded with black powder and sharp stones as projectiles. Black powder featured in the thirteenth-century writings of Arabic writer Abd Allah, Roger Bacon, and Count Albert of Bollstadt, Albertus Magnus. The first description of a primitive bomb comes from Markus the Greek in *The Book of Fires* from around the same time. "The second kind of flying fire is made in this way. . . . which these three things are very finely powdered on a marble slab. Then put as much powder as is desired into a case to make flying fire or thunder." How, when, and exactly who is responsible for the first black powder, and how it spread throughout Eurasia is nearly impossible to determine accurately. The evidence is vague, convoluted, and ultimately equivocal. The historian G. I. Brown writes in *The Big Bang*: "Any researcher is soon entangled in a web of mistakes, misinterpretations, and misrepresentations and the chance of finding a definitive answer has always been rather bleak." Nevertheless, it is clear that by the fourteenth century black powder was being experimented with throughout China, the Middle East, and Europe.

The earliest black powder, a blend of three rather mundane, naturally occurring substances, sulfur, charcoal, and saltpeter, was unreliable and difficult to use. The ingredients were generally impure and separated during transportation or storage, or easily became damp. Because it was uniformly ground, black powder could be packed too tightly into a weapon, slowing down combustion and causing misfires. Black powder burnt irregularly and clogged firearms after only a few shots. It produced tremendous quantities of smoke that blocked vision and ruined secrecy. Its use was an art as much as a skill, and required extensive practice rather than mere knowledge of proper handling procedures. The English gunner William Bourne explained in 1587 in his treatise *The Art of Shooting in Great Ordinance,* the tricky and fickle properties of working with early black powder, which was called serpentine because of its tortuous, unpredictable nature. "The powder rammed too hard," he observed, ". . . it will be long before the piece goes off. . . . The powder too loose will make the shot come short of the mark. . . . Put up the powder with the rammer head somewhat close, but bear it not too hard."

By the late sixteenth century, the quality and strength of black powder was greatly improved by the technique of corning. Pre-mixed powder was soaked in alcohol and water, dried, and then broken into crumbs of a uniform size that resulted in quick, reliable combustion. Corned powder was also more moisture resistant, and more suitable for use in the cannons of men-of-war and other armed ships. While the serpentine black powder was ground using a mortar and pestle, the improved corned powder was made using hand-operated stamping mills. One hundred years later, by the turn of the eighteenth century, great water-powered rolling mills remorselessly pulverized the alcohol-soaked powder and combined it into a dense substance known as mill cake or press cake. The press cake was mixed so thoroughly by the crushing that its components would not separate during transportation, and it was sieved to assure uniform grain size. The corned powder, using properly blended pure ingredients, was so effective that the recipe remained unchanged for more than three hundred years. The late-seventeenth-century writer and natural philosopher William Clarke wrote of the potency of corned powder that "since its first Invention to this very day, none could invent the like . . . a Fire so quick, vigorous, potent, dreadful, and not to be extinguish'd till wholly consumed."

Gunpowder's potency and power, however, comes from containment. When it is ignited by a flame (around 600 degrees Fahrenheit or 300 degrees centigrade) in the open, black powder quickly burns away in a cloud of smoke. Only when combustion is forced in one direction by containment, in a gun barrel for instance, do the expanding gasses rapidly build up enough pressure to launch a projectile, a round stone or iron ball, with the force to maim and kill, or to smash through wood and stone. Without the capacity to effectively trap, compress, and direct the power of the expanding gases, black powder's practical applications were limited. The era of gunpowder truly began when an effective recipe for black powder was combined with an effective tool for containing the explosion and launching a snugly fitting projectile. But what is simple in theory was difficult to achieve in practice. The origin of the earliest guns is as obscure as the origin of black powder.

Two crouching figures carved in a Buddhist cave in Szechuan Province in China appear to show the world's earliest black powder weapons. One clutches a smoking hand grenade and the other holds at his hip a fat glob-shaped gun with a blast shooting forth from the open end. Dating from the early twelfth century, these images have pushed the earliest use of guns back by nearly a century. The scarcity of references to these weapons indicates that they probably were not widely used or had not yet been perfected. The earliest guns in Arabia date from the early fourteenth century and were constructed out of bamboo tubes reinforced with iron bands. A charge of black powder propelled an arrow, probably less effectively than a trained bowman because of the irregular bore of the tube and the loose fit of the projectile. In 1327 the first image of a gun as an ungainly vase-shaped cannon appears in Europe in *On the Majesty, Wisdom, and Prudence of Kings* by the English soldier Walter de Milamete. By the fourteenth century, primitive cannons were being developed and used throughout feudal Europe. In 1331 a German army hauled cannons over a mountain pass into Italy. In 1346 English records show the purchase of black powder components for "the King's guns" before the Battle of Crécy.

These first primitive guns, odd-looking pear-shaped contraptions, consisted of bars of iron held together by hoops like a barrel. They produced a horrendous noise and great clouds of dirty smoke. They frequently exploded and killed or maimed the operators, causing more damage through fear and novelty than the shot. According to legend they were invented by Berthold Schwartz, a "German *Monk, or Chymical Philosopher,* who was a compleat qualified person in whom there was such a consociation of *Arts* and *Arms.*" "Black Berthold" reputedly lived in the city of Freiburg, which by the late fourteenth century was a center for gunnery training and forging. A fanciful engraving from several centuries after his death depicts him as a balding scholar garbed simply as a monk, closely observing an explosion erupting from a metal mortar in a primitive laboratory. Whether Berthold existed or not has not been reliably established, and many historians consider him to be a mythic figure, an amalgam and personification of the various technological developments in the vicinity of Freiburg. In the fifteenth century, the development of gunnery technology advanced rapidly, particularly in Europe, because of the technical skill of medieval bell makers at casting large bronze cathedral bells. Large cannon were cast using the same procedure.

Although cannons, or bombards as they were called, had caused rapid changes to the governing social structure of Europe by the end of the fifteenth century, small arms did not reach their pinnacle of military use until the early sixteenth century after the quality of gunpowder was perfected and gun design had improved. With the invention of a gun called the arquebus, using the matchlock mechanism, followed by the musket and regulated uniform shot size, guns became a standard feature of a soldier's outfit and arms. The demand for guns and hence black powder grew dramatically throughout the sixteenth century. By the time of Guy Fawkes and the Gunpowder Plot in 1605, and increasingly in the years that followed, securing a stable national supply of black powder was vital to the survival of European nations.

The recipe for gunpowder is simple and straightforward, but acquiring the ingredients in a pure form in quantities sufficient to meet the rapidly growing demand proved to be beyond the capacity of most nations. One of the key components of gunpowder was so valuable that wars were fought over it, and great commercial ventures were founded to scour the world for alternate natural sources. The hunt for this elusive, yet vital, substance led to some of the most bizarre and intrusive regulations and laws ever, and would eventually transcend the national boundaries of European nations as they exported their internecine quarrels around the globe. It was the beginning of a centuries-long quest for raw material that intensified with the development of new and more powerful explosives.

Black Powder's Soul
The Quest for the Elusive Saltpeter

It was a great pity, so it was,
This villainous saltpetre should be digg'd
Out of the bowels of the harmless earth,
Which many a good tall fellow had destroyed
So cowardly; and but for these vile guns,
He would himself have been a soldier.
—William Shakespeare, 1598
(Hotspur, *Henry IV,* Part 1)

The recipe "for making thunder and lightning" that Bacon tried to conceal so industriously and mysteriously is simple from a modern perspective: a blend of sulfur, charcoal, and saltpeter. The three components were ground into a powder, mixed thoroughly, and then sifted into a uniform mixture that varied in color from black to light brown depending on the proportion of charcoal. Black was the most common color and lent its name to black powder. A seventeenth-century philosopher named John Bate described the contributions of each of the components of gunpowder to the overall combustion process. "The saltpetre is the Soule," he claimed, "the Sulphur is the Life, and the Coales the Body of it." The saltpeter accelerates the burning of the sulfur, while the charcoal stabilizes the process and increases the combustible surface area. Speculating on how the explosion actually worked was the lofty role of natural philosophers; effectively making use of the gunpowder in artillery and bombs was the prosaic, yet respectable, job of gunners and soldiers; while the far less glamorous task of collecting and purifying the sulfur, charcoal, and saltpeter was carried out by an army of rural and urban laborers who toiled away in some of the earliest dirty industries of the preindustrial age.

The soft yellow substance known as sulfur, but also by its more colorful name brimstone (burning stone), is tasteless, odorless, and insoluble in water. It burns with a feeble blue flame at a low temperature that makes it safe for domestic use. It was used by humanity for thousands of years as a fire starter and for lamp wicks. Its acrid and noxious fumes made it valuable for incense in religious ceremonies, as an insect repellant, for pyrotechnic displays at festivals, circuses, and in theater, and for fumigating houses to cleanse the foul vapors that were believed to be the cause of most diseases. It was also used for bleaching textiles. It was a mysterious substance associated with the devil and hell, probably because in ancient times it was mined near volcanoes and natural hot springs, and because of its horrid stench.

The Bible mentions, amongst many other references, "brimstone and fire from the Lord in Heaven," and "Upon the wicked He shall rain Snares, Fire, and Brimstone, and a horrible tempest." The Greek poet Homer wrote, "Bring sulfur, old nurse, that cleanses all pollution and bring me fire, that I may purify the house with sulfur." The Roman historian Pliny (23–27 A.D.) reported that sulfur was a "most singular kind of earth and an agent of great power on other substances," and that it had "medicinal virtues." The English poets Milton, Southey, and Coleridge used it as a metaphor for the devil. Robert Southey wrote in "The Devil's Walk" that

From his brimstone bed, at break of day,
A-walking the devil is gone,

To look at his little snug farm of the World,
And see how his stock went on.

Although sulfur is common throughout the world, it was scarce in commercially viable quantities. It is often found blended with metals such as copper, iron, lead, and zinc or with nonmetallic elements such as barium, calcium (gypsum), sodium, or Epsom salts. Until the nineteenth century the primary location of easily mined sulfur deposits was in volcanic regions, particularly in Sicily, where a primitive industry evolved to render it, a polluting and wasteful process that destroyed much of the countryside. Workers heaped impure sulfur rock in a snakelike mound on a hillside and covered it with charcoal ash and dirt. They then ignited the buried sulfur and as the fire slowly moved down the hill the sulfur dissolved from the limestone and ran in great yellow stinking rivulets down to collecting vats. Nearly half the sulfur was consumed as fuel in order to purify the rest, and the acrid and poisonous fumes of sulfur dioxide soured the landscape for miles, acidifying and killing vegetation and giving the bleak devastated volcanic regions the appearance of hell on earth. The still crude sulfur was then refined a second time. It was boiled in an earthenware kettle, causing sulfur gas to vaporize and escape through a tube into a second container where it condensed, leaving the impurities behind.

Charcoal manufacturing was a task no more glamorous, the craft and trade having been long established by the time gunpowder became known in Europe in the fourteenth century. Charcoal, being almost pure carbon, burns at a very high temperature and produces little smoke. It was particularly desired as a household fuel and for heating rock to extract metals from ore. Creating charcoal from wood was dirty, hard work. The charcoal burner felled trees, sawed them into eight-foot bolts, and arranged them in twenty-foot-wide conical mounds around a central opening, like a dense teepee frame. They then crammed smaller pieces of wood into all cracks and shoveled dirt over the entire cone to help trap the heat. Burning charcoal was then placed down the central hole to ignite a fire that smoldered from the inside out until only charcoal remained. This primitive form of charcoal was not pure enough for strong gunpowder and as early as the late fourteenth century more efficient methods of making charcoal were developed, such as burning the wood in an iron cylinder. Different woods produced charcoal for use in different types of gunpowder: Hazelwood and dogwood were good for the small-grained rapid-burning gunpowder used in small guns, while willow and alder were more suited for the larger-grained powders used in cannons or for blasting rock.

The third and most important ingredient was saltpeter, the soul of gunpowder and all other future explosives. The term "saltpeter" comes from the Latin *sal petrae,* salt of stone, because it was first observed and collected as a white powdery efflorescence on brick and stone walls, or permeating rich soils. Arabic writers sometimes referred to it as Chinese snow. It was a much praised and invaluable substance throughout Europe because of its chemical properties, being used by alchemists as a separating and cleansing agent for metals, by glass manufacturers as a cleansing agent for sand, by textile dyers to set the dye, as a fertilizer, and as a preservative for meat. It was also used as a medicine because of its perceived value in curing a disparate and wide-ranging array of ailments. An early-seventeenth-century discourse written by an Irish gentleman named Thomas Chaloner to an "Apothecary dwelling at the signe of the Ewe and the golden Lambe over against Soper lane ende in cheepeside at London," listed many of the medicinal powers attributed to saltpeter, which were beneficial "to the use as well of the meaner people, as of the delicater sort." This list of seemingly incongruous uses of saltpeter included:

A lotion of washing to fordoe ringe-wormes, and tetters;

An oyntment or plaister to fordoe harde knobbes, buttons, or cornes whersoever they be in;

An other oyntment to fordoe the buttons of lepres faces;

An experte oyntment for cure of maunge;

A lotion or lynement for bringing out, and for outwarde healing of the small pocks, measelles;

For opening, riping, drawing and mendifying of carbuncles;

An other oyntment to fordoe Vermine and Nittes in the heade, and other hearie places;

To cleane teeth and for the tooth-ache;

To laxe and source the bowells.

Another book from the seventeenth century, *The Natural History of Nitre; or, a Philosophical Discourse of the Nature, Generation, Place and Artificial Extraction of Nitre, with Its Vertues and Uses,* by an English alchemist named William Clarke, provides a primitive analysis of saltpeter's chemical composition, and instructions detailing the places where it can be found, enlivened by meandering philosophical speculation and effusive praise of saltpeter's great power and brilliancy. He particularly emphasized saltpeter's role as an ingredient in black powder for "the great and Noble Art of Artillery." Saltpeter, he observed, "burns with *speed* and *vehemency,* not so gradually or mildly as other substances. . . . We may take notice of the *clearness* and *brightness* of its flame, dispersing itself into beams like the Sun in its greatest splendor." Clarke described saltpeter as nature's "hidden treasure" and he wondered "for what End and Use Nature hath lodged so excellent a Mineral in the Earth?"

Until the mid-nineteenth century saltpeter had but one source: the earth, primarily around farmyards where animal and vegetable remains decomposed. Saltpeter formed naturally in outhouses, barns, stables, and pigeon cotes. Enterprising collectors could find it in "crowded cities with narrow dirty streets and lanes where the decomposing organic matter with which the soil is impregnated becomes gradually nitrified, oozes through, and dries on the walls and floors of the cellars, as a whitish crust." Outhouses in towns were equipped with a special tray that was emptied and removed by the nightsoil collectors. On farms, people scraped the surface layers of earth and carted it off to the saltpeter beds for refinement. Clarke's advice to them was to "dig up some of the earth with a knife, and hold your hand till it be hot, which exerts its quality, and then taste it, if it be good it will prick the tongue, and taste like spice . . . the earth also will shoot and sparkle in the fire."

The curious process that caused saltpeter to form in one place and not another was a source of endless speculation for early European theorists anxious to understand the mysterious process that created this most invaluable of substances. "The laborious Bee," wrote Clarke, "makes the fabric of her Combs in the Dark, and within them the Quintessence of her honey. The Silk-worm works within her Web. The Physician's Experiments are privately dispensed, which, though freely communicated in their Use, yet are more concealed in their preparations: So that we know things themselves, but not the manner of their productions. No less obscure is *Nitre* in its Birth, and as difficult to be explained." Nevertheless, the author tries. "In the beginning of the World, it was first created in the Earth; and by the power given to it from the Creator, hath preserv'd and multiply'd itself. And so from the Creation, not only the formation of its [saltpeter's] own body, but its propagation and perpetuation proceeds."

Despite its unsanitary source, saltpeter was so valuable that alchemists like William Clarke attributed to it special, suitably lofty, origins. He dismissed the obvious observation that saltpeter was caused in some way by the "Parts, Urine, or Excrements of Men, Beasts, or Foul, the droppings of Wine or Beer, etc." and instead propounded that the illustrious substance was "attracted out of the air" and was the direct cause of meteors and lightning and thunder in the sky. Not to sully the reputation of such a valuable and pure substance he claimed that "what the Excrements or Urine of living creatures add, is not anything of their substance convertible into salt-petre, but their putrefactive heat . . . which heat and dryness are causes of its [saltpeter's] Breeding." So saltpeter was believed to be a product of the element of air, not the lowly earth. Despite his pretense to scientific analysis, Clarke's claims were merely a reflection, in the absence of any means to test and quantify scientific hypotheses, of the social and commercial value placed on saltpeter at the time. We now know that saltpeter is created by bacterial action during the decomposition of organic matter, particularly animal excreta. The process of nitrification is the conversion of nitrogen from animal and plant decay into nitrates, which accumulate in the soil. Nitrification is greatest where the concentration of nitrogenous organic matter is greatest, and it proceeds at a faster pace under higher temperatures that stimulate the bacterial action.

Once saltpeter-impregnated earth was identified and collected, it was brought to a leaching and boiling house where it was refined and purified. The same basic process was used for centuries. Various tracts—from as early as the 1500s to the process detailed by William Clarke in 1670 to the prosaic information supplied to farmers during the American Civil War—give layman's instructions for purifying saltpeter on a craft scale using niter beds. Into a shallow pit lined with clay, workers heaped saltpeter-infused earth in porous piles, about a meter deep and wide and six meters long, so that it resembled a giant burial mound of putrefying matter. (A field of niter beds could have been mistaken for a freshly dug graveyard.) In the seventeenth century this niter-producing mound was known as "concrete juice." Laborers then soaked the great odorous mounds weekly with liquid manure, urine, dung water, water from privies, cesspools, and drains. The earth was kept moist with these ripe fluids but not overly wet. As the mound "ripened," the saltpeter was evaporated to the surface as a whitish, saltlike powder. The top several inches of nitrified earth were scraped off the bed and brought to the leaching hut where it was shoveled into a large vat or barrel with a controlled drain on the bottom. The workers added water and frequently stirred the broth before allowing it to settle for a day and then drained it into a new container. This fluid was now known as the "raw liquor," and the leaching process was repeated several times to extract the maximum quantity of saltpeter from the bed before taking the "raw liquor" to the next stage of production.

The "raw liquor" solution was then boiled and mixed with wood ashes (potassium carbonate) or poured over ashes, which absorbed all the impurities, such as calcium or magnesium nitrates, and added potassium. The new solution, now known as "scoured liquor," was boiled yet again to remove the common salt and other remaining impurities. Because saltpeter is more soluble than common salt in boiling water, yet less soluble than salt in cold water, the boiling process concentrated the solution and allowed the common salt to crystallize as sediment, which was then scraped away from the bottom of the boiler. Impure organic scum also formed on the surface of the vat and was scooped away at intervals. The solution was then left to stand, and as it cooled the nitrate separated and crystalized on the bottom as a crude and impure form of saltpeter. The remaining solution was known as mother-of-Peter and was further refined by boiling again and adding glue or blood, which seized upon the remaining organic matter and formed more scum, which was scraped away until the liquid was clear. The final cooling produced commercially viable saltpeter crystals. Saltpeter production was a time-consuming, painstaking, messy, and laborious process.

To many the making of saltpeter was as much an art as a trade and the inspiration for some ornate and effusive prose by those involved in its manufacture. "I know not what Experiment I have taken more pleasure in," wrote the colorful William Clarke, "than to free such a *Crystalline Substance* taken out of *dirty Earth*, so *beautiful* a *Body* out of *Chaos*, such a *Spiritual Essence*; as without a metaphor almost it may be called, drawn from a *Caput Mortuum* . . . the Mechanical Labourers at the work cannot behold it without admiration." But it wasn't just admiration and love of beauty that inspired the saltpeter makers of seventeenth-century Europe, it was something much more prosaic—national security.

Despite the great effort and labor required to produce even small quantities of saltpeter, there was no alternative. The manure- and urine-spattered earth from a moderate-sized stable, if managed properly, could yield with each leaching about one thousand pounds of the coveted saltpeter. But even at this rate of return it was nearly impossible for European nations to obtain a sufficient supply during times of war—and as the use of gunpowder in guns, cannons, bombs, and even in mining increased, the demand for saltpeter also increased. For every one part of charcoal and every one part of sulfur, military-grade gunpowder required six parts of saltpeter. The continent could not produce the quantities it needed to wage its wars, fertilize its crops, and experiment with new techniques in blasting. In the early seventeenth century, nations across Europe realized that they didn't have enough of the powdery efflorescence to pursue their ambitions and defend themselves from the encroachments of their scheming neighbors.

In 1626, King Charles I of England made an astonishing pronouncement, surely one of the most unusual proclamations of the

A sixteenth century engraving shows the primitive preindustrial method of breeding saltpeter in "nitre beds" for gunpowder.

era, or perhaps of any other. Addressed to his "loving subjects, inhabiting within every city, town and village," it was a command to "carefully and constantly keep and preserve in some convenient vessels or receptacles fit for the purpose, all the urine of man during the whole year, and all the stale of beasts which they can save . . . and that they be careful to use the best means of gathering together and preserving the urine and stale." The people's urine was to be poured on the niter beds to breed saltpeter. The king and his advisors believed the order was "necessary for the public service" because of the need for saltpeter for gunpowder. Only persons of "quality" were exempted, and dissenters would be punished for being "contemptuous and ill-affected both to our person and estate," and punished "with what severity we may." Although the intrusive order remained in effect for only one year, other proclamations, not nearly as odious but perhaps as frustrating, were to annoy the English citizens for much of the seventeenth century, and the citizens of other European nations as well.

Two years earlier, Charles I had issued a proclamation admonishing his subjects for impeding the natural growth of saltpeter by paving the floors of their homes and barns and making wooden walkways between buildings. It was, according to the king, an example of the people placing their selfish personal interests ahead of the public good and the nation by destroying saltpeter production. "Whereas the making of Salt-Peeter and Gunne-powder within Our realms and Dominions, is a great benefit, strength, safety and defence to Us and Our Subjects in generall . . ." began the proclamation, saltpeter is "not to be obtained but at the pleasure of other Princes, at unreasonable rates and prices, and the same being so obtained, may happen to be intercepted, or by contrary winds hindered, or bitterly lost by shipwreck, or such like Casualties on the Sea . . . [and] whereby the Treasure of Our Realmes would be much diminished, and other foreine Realms thereby enriched. . . ." The proclamation called for "the preservation of Grounds for making of Salt-Peeter, and to restore such Grounds which now are destroyed, and to command assistance be given to his Magesty's Salt-Peeter-makers." In this and other similar proclamations in the following decades, Charles I commanded that interior floors of houses and barns were to "lie open with good and mellow earth, apt to breed an increase of the said saltpeter," thereby legally mandating what would today be considered a violation of the public health code.

Not only did the public-minded Charles I order citizens to maintain dirt floors and mud pathways, he also broadened the powers of a public officer who, although only in existence for several decades, was more reviled and feared than the tax man, and far more unusual—the saltpeterman. As early as 1606, a commission was established to look at the abuses of property by the overzealous industry of the saltpetermen, or petermen, complaints arising soon after they were unleashed onto the countryside in the late-sixteenth-century reign of Elizabeth I. The petermen held a warrant from the crown, or merchant supplier to the crown, to scour the countryside for likely sources of the treasure—privies, pigeon cotes, and dunghills, and to break up the earth and dig for it and haul it away to be processed. According to the numerous complaints heard by various justices over three decades, the petermen did not endear themselves to the people. They took bribes, they arrogantly commandeered carts for the transport of the impregnated earth, they ruthlessly tore up the floors of dwelling places and barns, leaving the planks broken or chaotically heaped afterward. They threatened people with extra digging if they failed to immediately comply with their demands or complained; they dug the floors of barns during harvest, burrowed in pigeon coops during breeding season, and scraped away in malthouses in the midst of the fermenting. In short, they came when they wanted without regard to the inconvenience to the people whose homes and business they invaded. According to Sir Thomas Jervoise, who forwarded the complaints aired at several public assemblies to the commissioners for the Admiralty, who were overseeing the matter in 1629, the reviled petermen even broke up the floors "in bedchambers placing their tubs by the bedside of the old and sick, even women in childbed and persons on their death bed." In another instance, it was reported that "if any oppose them they break up men's houses and dig by force." Although Charles I did acknowledge "the trouble and grievance of our loving subjects by digging up their dwelling houses," he was determined to reduce England's reliance on uncertain foreign saltpeter supplies, which typically came from the fertile regions of Barbary, France, Poland, or Germany, or were unavailable at any price during war.

The historian Kevin Sharpe noted in his comprehensive biography *The Personal Rule of Charles I*, "In general the policing of saltpetermen was impossible. Gunpowder was urgently needed. As Viscount Wimbledon observed in 1635, 'it were better the Kingdom of England were without walls than powder . . . in time of peace provision must be made for war, for in time of war it is too late.'" The ravages and intrusions of the petermen were not curtailed until 1656, when an act of Parliament required the petermen to obtain landowners' consent before scraping the earth on their premises. The industrious striving of the petermen, however, could not keep England supplied with even one third of her total need. In 1639, Charles I was obliged to import saltpeter to arm his troops for his campaign against Scotland. A late-seventeenth-century public petition "humbly offered to the consideration of the Honourable House of Commons" outlined the problems caused by a saltpeter shortage. "Saltpeter is a Commodity so necessary, that we can neither Fight nor Trade, or send a ship to sea, without it; and in time of war especially, it is our Interest to buy it, if it cannot be procured otherwise, of our Enemies. Therefore, upon the present occasion it is not material how much those we must buy it of will get by us; but whether we want it or no."

The petition also sheds light on the many uses for saltpeter in that era. "By reason of the scarcity of saltpeter in England, which is a great ingredient in the composition of Glass, the Glass men are nigh ruined. Our Scarlet and Bow Dyers cannot Colour without saltpeter; and because of the price it now bears, merchants . . . send their Cloth to be dyed in Holland; and the Dutch have a great advantage of us in dying their own clothes cheap for the *Turkey* Trade, to the prejudice of the Woollen

Manufacture of this Nation." The local "Pattentees have been able hitherto to make but a very inconsiderable quantity, two Tun at most, tho in respect of the great price this commodity bears, it was in their Interest to make as much as was possible."

It was not only England that was desperate for saltpeter, but all of Europe, and the demand grew intense when nations were at war or threatening war. The office of the saltpeterman in one form or another existed throughout Europe (*salpetriers* in France), where in general any soil impregnated with saltpeter was considered to be the property of the crown. Throughout Europe no farm, stable, pigeon cote, slaughterhouse, privy, or public house was safe from the roving bands, who earned a commission based on the quantities they could secure. Without saltpeter for gunpowder, armies and ships had to rely on outdated handheld weapons. "The Navigator," lamented William Clarke at the time, "might better sail in the Vast Ocean to his desired Port without the *Needle* touch'd with the *loadstone* in the *compass,* than the *souldier* do any Execution without this *powder:* his Great Guns would be but dead bodies, and no more dreadful, were they not animated with this *Nitro-Sulphureous Spirit.*" Saltpeter was a strategic raw material of warfare. All of Europe was desperate for the limited supply, but there was never enough. France routinely produced only half of her required supplies, and England only a third.

Fortunately for the quarreling nations of Europe a new and much greater source of the coveted saltpeter was discovered in the early seventeenth century.

The sun was high overhead and the air was burning and dry. Through the dust and heat a curious procession of dark-skinned men clad in white loincloths shepherded a reluctant herd of lazy cattle toward a great swath of churned, rich earth at the farthest corner of a farmyard. With shouts and sticks they kept the docile beasts hemmed onto the soil for half an hour before leading them back to the fields where they grazed. Throughout the day they brought a series of cattle herds to the field and kept them milling about the earth for about half an hour before leading them away. Later that day they brought the same herd for a second time, repeating the process throughout the day. Between herding, the men leaped onto the damp earth and turned the soil over with shovels and hoes until it slowly became sodden with the urine of the cattle.

The workers were the low-caste Nunias and Beldars, the workers who labored with hoes in the earth, digging ditches and making roads. Their prime occupation during the dry season was the making of crude saltpeter to meet the growing demands of the Dutch and English companies with their great warehouses and offices in Patna, upriver along the Ganges in the northeast of India. By October the rich soil began to produce the crystal efflorescence of raw saltpeter, which the Nunias scraped and dragged away for boiling and refining. Secrecy was paramount in their tedious undertaking with the cattle, not because it was illegal but, according to the historian Narayan Prasad Singh, "in order to avoid offense of affluent people. . . . The people who had most cattle being either pure Hindus or Muslims of rank, had an aversion to allow this operation, as they either abominated the Beldars and Nunias as impure or were jealous of their prying near the women [in the farmhouse]." Although the saltpeter collectors would have preferred to wander about the farmyard scraping the earth in the choicest locations, they were compelled to undertake the time-consuming and inefficient job of herding cattle onto a distant plot because they were forbidden to "roam about in the premises of the residential areas of the affluent people of the society where cow herds were kept. . . . Their contact with the womenfolk of the affluent families was studiously avoided, for it was feared to bring odium to those coming in their contact." Nearly all the people of a village would permit their cattle to be herded onto the distant saltpeter plots rather than allow the men to work close to their homes and families.

The untold labor and sweat of the Nunias and Beldars earned them little remuneration. It was a dirty, unrewarding job. But they produced such a high quality of saltpeter that by the end of the seventeenth century it was the primary supply for most of Europe, and a source of great wealth to the middlemen, brokers, and government officials who regulated the trade. Saltpeter bred with particular vigor in the sewage-sodden soils of the agricultural heartland of Bihar and Bengal, where the extraordinary heat and prolonged dry season produced great quantities of the highest quality. "East India," according to one seventeenth-century observer, "gloryeth as much in this [saltpeter] as in its spices." Despite a very primitive and inefficient craft or artisan organization of the industry, the quantity of saltpeter produced in Bihar was vast. The entire region was "a veritable Eldorado for the saltpeter business," and surpassed all others in quantity, quality, and price. By the eighteenth century, all the European companies had an agent, warehouses, and social or commercial relationships with the various producers. Other saltpeter-producing regions of India were soon abandoned because they could not guarantee the quantities needed for European ships, which, because of its weight, used saltpeter as ballast, before they set sail with their other valuable cargo. Indian saltpeter fueled to a large extent most of the European wars from the mid-seventeenth century through the eighteenth century.

The second half of the seventeenth century and the early eighteenth century saw a continuous series of conflicts in Europe involving Sweden, Denmark, France, the Netherlands, Spain, Portugal, the Holy Roman Empire, Russia, Poland, and the Ottoman Empire in an endlessly shifting round of alliances. There

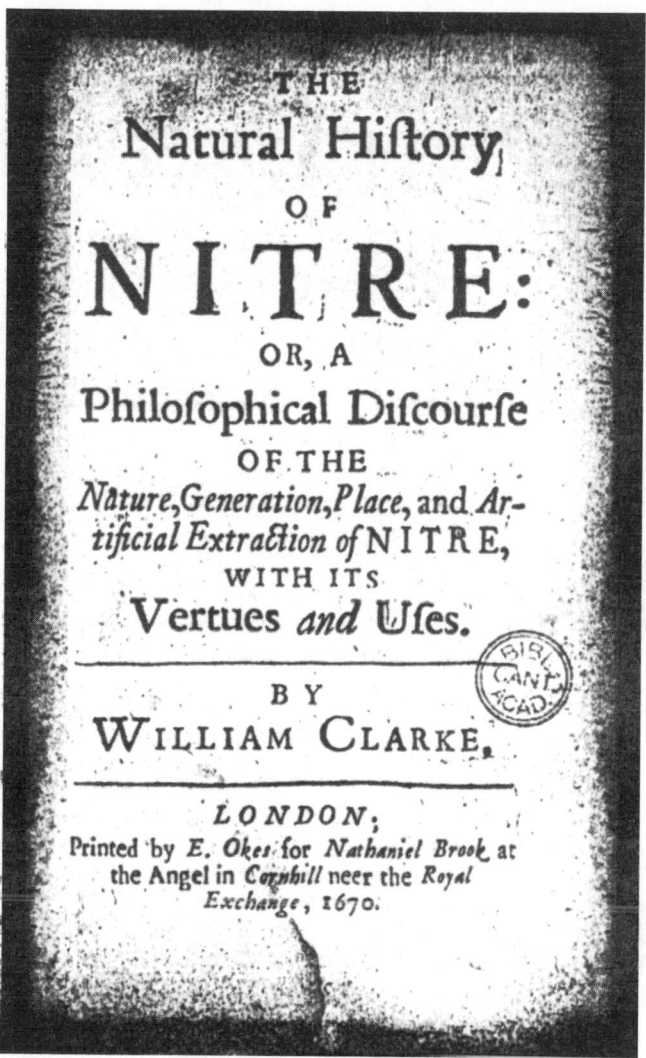

This is the title page of William Clarke's A Natural History of Nitre, *ca. 1670. Niter, or saltpeter, is necessary for the creation of gunpowder and all other explosives. It was one of the most coveted and scarce substances in seventeenth century Europe and one of the key commodities returned from India as late as the mid-nineteenth century.*

was scarcely a handful of years when a war was not being fought somewhere on the continent. As a result, the price and demand for saltpeter varied wildly, and it was never a consistently profitable commodity despite the occasional dire need and shortage. It was called "a bad neighbour to better goods" like indigo, silk, and the fine cotton known as muslin. Although the demand fluctuated wildly from year to year, it steadily increased throughout the century. The wild and chaotic fluctuations of the demand and price of saltpeter throughout the period has been analyzed by the economic historian K. N. Chaudhuri in his comprehensive study *The Trading World of Asia and the English East India Company, 1660–1760*. He concluded that "the most important factor regulating the supplies of the demand side was the condition of war and peace in Europe." Holden Furber similarly writes in his book *Rival Empires of Trade in the Orient, 1600–1800* that throughout the second half of the seventeenth century "the company's sales, with their steadily rising receipts from Bengal saltpeter, reflected an ever more warlike Europe."

Portuguese traders had been quietly shipping Indian saltpeter from their base at Goa for most of the sixteenth century, but the quantities were often small and were consumed in their domestic market. The first traders to contract for saltpeter for export to the European market were the Dutch mariners of the Dutch East India Company, who had established factories and warehouses along the Coramandel coast in the southeast early in the seventeenth century. English merchants were not long behind, establishing factories at Madras and on the west coast at Bombay (after being repelled from the Indonesian spice trade by the Dutch). Saltpeter quickly became one of the primary export commodities for both companies. The historian Jagadish Narayan Sarkar observed in the *Indian Historical Quarterly* that "saltpeter was so much in demand in England that there was a standing order from the Company's authorities there for an annual supply." In spite of the wild price fluctuations, English and Dutch companies reaped vast profits from their mercantile activities, and paid huge dividends to their shareholders and taxes and duties to their respective governments.

Competition intensified in the early eighteenth century. In addition to the Dutch and English East India Companies, there were French, Danish, Swedish, and Austrian companies, all vying for Indian saltpeter. The Dutch dominated in all ways, with the largest warehouses, the most experienced people, and the most efficient barge transport system (saltpeter was too heavy for overland transport). English factors, or company agents, in the early days recorded their predicament with a touch of jealousy: "The Dutch manage things better," wrote one wistful factor, while another complained that "the Dutch are insolent and feare not to break all contracts." The politics of securing saltpeter from Bihar was like an elaborate and intricate ballroom dance, with changing tune and tempo, shifting partners and alliances as they swirled about the great hall. Lurking beneath the stately deportment of the dancers were their petty jealousies, intrigues, and secret agendas. Subtle manipulations and crass threats, bribery and blackmail, and occasionally violent action were all part of this dangerous and high-stakes ritual that played on for a century. In order to secure regular shipments of saltpeter, company agents had to become intimately knowledgeable and involved in the politics of the region: They had to know to whom was owed the official tax, whom to quietly bribe, and to whom to complain when problems arose, as they invariably did. After decades of business presence, the traders had deep political and social connections within government and the leading merchant families.

The emperor of the Mughal Empire, which governed most of what is now India, claimed as a royal prerogative a monopoly of all saltpeter in the region, and occasionally all other goods and trade as well. The historian Narayan Prasad Singh discusses the sometimes rocky relationship between the government and the European traders. In 1646 a local prince, Aurangzeb, who later became the emperor, prohibited the sale of saltpeter to Christians because "the Prince (very superstitious) possessed by some of his churchmen that it is not lawful for him, to suffer us to export that specie which peradventure may be employed against Moores." Corruption was rife and securing saltpeter often required bribery, blackmail, and other creative schemes to placate the whims and desires of the governors, officials, and functionaries, and later the powerful Indian merchant families. One English factor wrote in 1664 that the local governor Shaista Khan "demands English and Dutch men to serve him in said warrs, and doth expect to be furnished." A factor named Job Charnock wrote in 1679 that "here is noe order of Government. Every petty officer makes prey of us and abuses us at pleasure to screw what he can out of us." While Mughal officials conspired to drive up the price of saltpeter through customs duties and collusion, European companies tried to drive the price down through joint bargaining, but the efforts usually failed because they bickered and quarreled among themselves.

In 1739, an armed band of Dutch traders seized a shipment of saltpeter that had been jointly purchased by the English and themselves, and only after complaints to the local governor, and probably bribes, was the shipment ordered to be properly divided. In 1746, Dutch factors in Patna, through their influence with the barge owners, prearranged the hiring of all the barges for their own huge shipment of saltpeter, leaving the entire English shipment stranded inland. During the 1740s and 1750s, Dutch traders controlled nearly half the entire saltpeter production of Bihar, far outstripping any other nation.

While the Dutch company hauled tons of saltpeter back to Amsterdam and promptly exported it throughout Europe, the

English company did not have the same freedom to exercise its mercantile inclinations. It was obliged to bring the saltpeter back to England as a public relations gesture, to assuage the ever present agitators who demanded an end to its monopoly privileges. In 1702 the English company, as part of its new charter, was required to supply the English Crown with five hundred tons of saltpeter annually at a favorable price or lose the right to export silver bullion from England, the currency of the eastern trade, without hefty export duties. In a sense, the company paid the English crown saltpeter to maintain its profitable monopoly. Unlike the Dutch company, whose prime objective was profit, the English company never operated entirely as a trading enterprise. When England was at war, the company was forbidden to export saltpeter to the continent through its normal channels, so it actually sold more saltpeter during peacetime. Its dealings in saltpeter were always closely linked with Britain's political aspirations because of the precarious state of the English saltpeter industry—a situation that led the company into ever greater involvement in Indian political affairs in an attempt to stablize its saltpeter supply.

In 1705, the Mughal emperor Aurangzeb died at the age of eighty-eight, and in the years following his death the power of the Mughal Empire steadily declined, and commerce became ever more challenging and difficult. The Mughal dynasty was descended from the Mongols, an invading force that swept down into India from Central Asia in the sixteenth century, soon after Portuguese traders established their outpost at Goa. Throughout the sixteenth century, Mughal armies marched and conquered, slowly extending their rule over most of the Indian subcontinent, and what is today Pakistan and parts of Afghanistan. Aurangzeb, who ruled from 1658 to 1705, was a religious fanatic and bigoted Muslim who persecuted Hindus throughout his realm. When he died, his empire began to disintegrate, as local rulers who had chafed under his heavy-handed rule seized the opportunity to assert their independence. As authority waned, the central government was increasingly unable to maintain the peace, and travel and trade became more subject to the whims of local lords and bandits. Corruption ballooned as the hierarchy disintegrated. In the absence of a strong government the companies began to arm themselves and maintain small professional standing armies that they hired out to local rulers to settle regional power struggles.

The companies sought to secure political and economic advantages by manipulating political and economic affairs. After years of increasingly acrimonious squabbling, the French company, under Joseph François Dupleix, sought to establish European political control upon the ruined foundation of the Mughal Empire. In 1746, aided by French troops sent to India to harass the British during the War of the Austrian Succession, he attacked and seized the town of Madras and sought to expel the English from the region. Bitter fighting between the two companies was aided by troops from Europe and from their respective Indian allies. The war ended in 1749, and Madras was returned to the British. The short conflict revealed the superior weapons and training of the French and English soldiers, and the companies began to be seen in a new light—not merely as innocuous traders but as formidable military powers. Despite the official peace in Europe, the conflict in India between the companies continued, with the French and the English backing rival claimants to be the ruler, or nawab, of Arcot (the region surrounding Madras in southern India) and supplying them with troops and weapons. When the Seven Years War began in Europe in 1756, it quickly spilled over into India, intensifying the ongoing struggle between the two companies. The eventual outcome was that Robert Clive led the British forces and their allies to victory over the French and their allies, ending French power in India. In 1757, at the Battle of Plassey, Clive's 1,100 British soldiers and 2,100 Hindu sepoys defeated the nawab of Bengal's 50,000

Mughal foot soldiers and horsemen, and later a Dutch force at Chinsura, establishing the beginning of British rule in India and cutting off shipments of saltpeter to the French until the war ended in 1763. The lack of saltpeter for gunpowder has been proposed by historians as one of the key factors influencing the French to sue for peace in 1763 at the end of the Seven Years War. In 1765, the Mughal emperor Shah Alam II granted the English company administrative control over the provinces of Bengal, Bihar, and Orissa, transforming the company into an imperial power.

After conquering Bengal, the British were quick to check the Dutch dominance of the saltpeter industry, imposing strict quotas on the substance for all the European companies. The Dutch were at first allowed 28,000 maunds of saltpeter annually (less than half of their export during key years of high demand in Europe), which was decreased to 23,000 maunds in the 1760s. There was an allocation of 18,000 maunds for the French, and 16,000 maunds each for Denmark and the United States, after it entered the market in the early nineteenth century. One maund is equivalent to about 82 pounds. Under British control, access to the saltpeter was cut off entirely to hostile nations in times of war and occasionally cut off to neutral nations when the British feared it would be sold to their enemies. Saltpeter became increasingly a tool of power politics. France's sudden loss of Indian saltpeter in 1758, and her forced reliance on poor-quality domestic powder, at least partly accounts for the disastrous defeat she suffered in the Seven Years War. If not for the competence of Antoine-Laurent Lavoisier, who as head of the Régie des Poudres beginning in 1776 rejuvenated the domestic saltpeter industry by dramatically boosting domestic production and improving the quality of French gunpowder, France would have had limited firepower against Britain during the War of American Independence. (There was no local production of saltpeter or gunpowder in the American states at the start of the revolution, and Britain naturally suspended shipments. A secret committee headed by Benjamin Franklin arranged for the purchase of gunpowder from France and the Netherlands.) During the French Revolution and Napoleonic Wars the British trade blockade left France entirely dependent upon more expensive domestic supplies of saltpeter throughout the course of the conflict. During the War of 1812, the Royal Navy blockade of U.S. ports shut off the supply of Indian saltpeter, stimulating another frantic search for the vital substance, and during the Civil War the North was dependent on British saltpeter from India.

The British acquired a stable supply of saltpeter for themselves and a powerful weapon against their enemies when their relationship with Bihar and Bengal shifted from trade to government. Although the British dominated the industry, Indian saltpeter remained the prime global source of the vital substance well into the nineteenth century. The economic historian Holden Furber has documented the dramatic increase in saltpeter consumption throughout the eighteenth century, concluding that "demand rose not only because the wars of the seventeenth and eighteenth centuries required more powder than those of the sixteenth, but because improvements in manufacture called for a higher ratio of saltpeter to sulphur and charcoal. Between 1660 and 1785 the proportion of saltpeter in gunpowder steadily went up from 66 percent to 75 percent; it had averaged 50 percent in the sixteenth century." The rising quantity of English saltpeter exports from India is revealing: In the 1660s, six hundred tons of saltpeter were shipped annually; by the War of the Spanish Succession in the early years of the eighteenth century it had risen to two thousand tons annually; and during the Napoleonic Wars at the end of the eighteenth century the annual shipments had risen tenfold, to an incredible twenty thousand tons.

Eventually, even India's thriving saltpeter industry was unable to meet the increased demand for saltpeter that was stimulated by several mid-nineteenth-century scientific and technological advances. A new and even greater source of the vital substance was soon discovered, which enabled this new and powerful explosives technology to quickly revolutionize the western world.

Blasting Oil and the Blasting Cap

Alfred Nobel and the Terrible Power of Nitroglycerin

Well, it is fiendish things we are working on, but they are so very interesting as purely technical problems and . . . clear of all financial and commercial considerations, they are doubly fascinating.
—Alfred Nobel, ca. 1867

On the brisk morning of September 3, 1864, a sickly Swedish chemist was on his way to Stockholm to meet the wealthy industrialist J. W. Smitt about securing investment financing for his latest business venture. When he departed the ramshackle shed in Heleneborg that he optimistically called his laboratory, his younger brother Oscar-Emil and another young chemist named Carl Eric Hertzman were busy producing a large quantity of a new explosive oil that was soon to be shipped to the State Railway crews who were laboriously blasting tunnels through hard rock with gunpowder en route to Stockholm. It was an exciting time for the thirty-year-old inventor and entrepreneur, and all he needed was capital to expand his operation. His new blasting oil was far superior to gunpowder, perhaps five to ten times as powerful, and perfectly safe too, he believed, if produced and handled properly, and detonated according to the specific method he had outlined in his recent patent application.

While he was in Stockholm there was a terrific explosion in Heleneborg. A blast destroyed most of the laboratory, and flames quickly began consuming the structure. Fire shot forth from the windows and door and began to spread to nearby buildings. Glass was shattered throughout the compound. Neighbors rushed from their homes and hastily formed a line with buckets of water and halted the fire's advance. When it was finally doused, little remained of the laboratory. A crowd of frightened onlookers gathered near the smoldering ruins and dragged from the blackened timbers and debris five charred bodies—a random passerby, the family's maid, and a local handyman, in addition to the two young chemists. By the time the young inventor rushed back from Stockholm there wasn't much left to do. His parents were in their modest house nearby, and he went to hear the account of the terrible news. The father of the two brothers, one killed instantly, the other left to ponder the events, later commented: "As no one survived I can only infer, from some remarks made by my dead son before the disaster, that the explosion was due to an attempt he made to simplify the method of producing explosive oil." At the police inquiry authorities were stunned to discover that Immanuel Nobel and his two sons had been illegally manufacturing explosives within city limits.

The awful setback occurred just as Alfred Nobel was perfecting the technical components of his most significant scientific discovery, a discovery that has since been hailed as the single greatest advance in explosives technology since the invention of gunpowder, a discovery that would have profound implications for a rapidly industrializing Europe and would fundamentally redirect explosives research forever.

The substance that caused the explosion in Nobel's small laboratory was nitroglycerin, a volatile and poisonous liquid that had been discovered sixteen years earlier by Italian chemist Ascanio Sobrero while he worked at the University of Turin. The early nineteenth century was an era of great experimentation in the relatively new science of chemistry. Chemists were frequently discovering new substances by trial and error and eagerly testing their properties in search of something useful. In the 1830s, a French scientist named Théophile Pelouze was working with nitric acid and discovered its explosive properties. Sobrero first developed his interest in nitric acid and its effects on organic substances as Pelouze's student in his private laboratory in Paris between 1840 and 1843, and he continued this research when he returned to Turin. He was appointed a professor of applied chemistry in 1845 and built a modest laboratory for his experiments. His experiments became dangerous in 1846 when he mixed nitric acid and sulfuric acid with glycerin, an inert by-product of the manufacture of soap. Sulfuric acid was created by burning sulfur and potassium nitrate (saltpeter) and allowing the fumes to be absorbed into water. It was originally called oil of vitriol in the eighteenth century and was promoted by an English quack doctor named Joshua Ward as a cure for scurvy, among other ailments. Nitric acid was created by heating sulfuric acid with additional saltpeter. It was used as early as the sixteenth century by alchemists and was called aqua fortis, strong water, because it could dissolve all metals except gold. It was used to separate gold from silver.

Sobrero mixed one part glycerin with two parts sulfuric acid and one part nitric acid at a temperature below the freezing point of water, and when the reaction was complete, he poured the solution into a demijohn of water. The nitroglycerin sank to the bottom as an oily pale yellow liquid like salad oil. It looked harmless, but it wasn't. "A drop was heated in a test tube," Sobrero related, "and exploded with such violence that the glass splinters cut deep into my face and hands, and hurt other people who were some distance off in the room." He reported his discovery in a letter to Pelouze, in February 1847, as recorded in George MacDonald's *Historical Papers on Modern Explosives*. Nitroglycerin also proved to be poisonous. When Sobrero, in the cavalier manner by which experiments were conducted in those days, placed a small drop on his tongue, he immediately knew he had made a mistake. It caused a "pulsating, violent headache accompanied by a great weakness of the limbs." It took him several hours to recover from even this minor quantity of the substance. He wisely confined future experiments to rats and dogs and other lab animals. When he fed a dog a small spoonful of the oil the poor beast "began to foam at the mouth and then vomited" and for several hours lay "trembling violently, and beating its head on the wall" before dying. "It is advisable," Sobrero wrote with great understatement, "to take great care in testing this property."

Further studies revealed the substance to be very capricious. Sometimes it exploded unexpectedly, sometimes when lit it merely burned away slowly, leaving an oily residue. It corroded metals, making it very difficult to store. Its volatility and power made it dangerous to manufacture. Because it was five or more times as powerful as gunpowder, however, it inspired experimental chemists throughout Europe to search for a method to harness its energy for use in firearms and explosives. Around the same time that Sobrero discovered nitroglycerin, a Swiss chemist, Christian Friedrich Schönbein, produced a similar substance that he called guncotton, after soaking cotton wool in a mixture of sulfuric acid and nitric acid. It, too, proved highly explosive but impossible to control. Devastating explosions killed dozens of people within a very short time, and guncotton, like nitroglycerin, was at least temporarily abandoned. Despite its obvious practical uses and earth-shattering power, it took years before nitroglycerin and other similar creations evolved into more than scientific curiosities. Scientists could not find a means to safely manufacture nitroglycerin or to safely and predictably detonate it. It could not be tamed. Sobrero wrote that "it is not yet possible to say anything as to the use that may some day be found for this liquid substance, which can be exploded by a shock; future experience alone will show us."

———

The man who would eventually tame nitroglycerin enough to market it and sell it was a natural inventor, born into a family of inventors and entrepreneurs. His father, Immanuel, was a creative and optimistic dreamer, but only marginally successful financially, and the family had to weather wild fluctuations in their fortunes. After a fire destroyed his factory in the 1830s, forcing him into bankruptcy, he fled Sweden for Russia, leaving his wife, Andriette, and sons, and his creditors, behind for several years while he strove to reestablish himself. The family scraped out a living running a small milk and produce store on the outskirts of Stockholm, while the eldest son, Ludvig, aged seven, sold matchsticks to earn money for dinner. In St. Petersburg, Immanuel cajoled the Russian military authorities into supporting his wild scheme for the development of naval mines, and by 1842 he was a man of power and influence, having parlayed his ideas into a flourishing industrial empire employing over a thousand people (a great number for the time in Russia, which was technologically lagging behind the rest of Europe). His business manufactured armaments for the Russian government—mines, cannonballs, and mortars, and eventually other iron components for steam engines and industrial machines. Finally, after repaying all his Swedish creditors, he sent for his family to join him.

Alfred Nobel was nine years old when his family moved to Russia, and it was here that he received the only formal education of his life under the professional tutorship of two Russian teachers, professors Nikolai Zinin and B. Lars Santesson. Although he was withdrawn and seldom played with other children, he proved an excellent student. His father wrote that "my good and industrious Alfred . . . is held in high esteem by his parents and brothers, both for his knowledge and for his untiring capacity for work, which is excelled by no one." He particularly excelled at languages and eventually became fluent in Russian, German, English, French, and Italian, in addition to his native Swedish. Encouraged by his mother, he developed a lifelong love of poetry, despite being discouraged by his father, who considered it frivolous, and eventually wrote his own skillful but imitative poetry in English. Chemistry he mostly studied on his own to satisfy his curiosity. Physically, however, the boy was sickly, scrawny, weak, and prone to illness—as he would be all his life. One of his own poems sheds light on the impact of his frequent illnesses, which kept him bedridden for great periods of time.

When fellow boys are playing
He joins them not, a looker-on,
And thus debarred the pleasures of his age
His mind keeps brooding over those to come.

In 1850, at the age of seventeen, Alfred was sent on a grand tour to study abroad and, his father hoped, to shake him out of his quiet introversion. For two years he traveled between Germany, France, Italy, and the United States, where he met with the famous Swedish-American engineer John Ericsson. He particularly was drawn to Paris, where he spent time working in a chemical laboratory.

At the start of the Crimean War in 1853, Alfred rejoined his family in St. Petersburg to work in the rapidly expanding factory, which was churning out munitions for the Russian army. One of their products was Immanuel's innovative underwater mines, crude watertight wooden casks stuffed with gunpowder, which were submerged on a chain across the entrance to the Gulf of Finland. Astonishingly for a new idea that underwent little testing, they served their purpose and exploded when hit by ships, significantly damaging a squadron of British and French ships in June 1855. The Nobel industry prospered throughout the war, but when hostilities ended in 1856 the military contracts were abruptly canceled, driving the business into bankruptcy soon afterward. By 1858 the elder Nobel, his wife, and their youngest son, Oscar-Emil, were back in Stockholm. Ludvig, Robert, and Alfred remained in St. Petersburg to wind down the business and salvage what they could of the family's reputation and the business assets. (Robert and Ludvig eventually were so successful in manufacturing munitions that in the 1870s they branched out into crude oil production in the Baku region and made a fortune.) While in St. Petersburg, Immanuel and Alfred were introduced to nitroglycerin by Alfred's tutor, Zinin, and professor, Yuli Trapp, who demonstrated its properties by putting a drop on an anvil and smacking it with a hammer, causing a bang and a flash. Alfred was enthralled. Immanuel tried for years to control the erratic, wild substance for possible use in his sea mines, but he never discovered a reliable means of igniting it and eventually abandoned his experiments. Both father and son, however, never gave up their interest in the volatile liquid, intuitively knowing that to harness its mighty potential would secure the family's fortune.

Back in Heleneborg in the early 1860s, Immanuel began experimenting with mixing nitroglycerin and gunpowder. Meanwhile, in St. Petersburg, Alfred was working along similar lines, but rather than mixing the two substances, he was approaching something truly revolutionary—using one substance to ignite another, the gunpowder to deliver the initial shock that would detonate the more powerful nitroglycerin. His first experiment was in St. Petersburg in a drainage ditch near Ludvig's factory. In May 1862 he poured nitroglycerin into a glass cylinder, sealed it, and placed the glass cylinder inside a larger metal cylinder that he then filled with gunpowder. He sealed the larger cylinder, except for a fuse. With his two older brothers looking on, he lit the fuse and tossed the infernal contraption into the canal where it exploded with tremendous force, sending a geyser of dirty water skyward. This experiment was a success, but there were many imperfections to work out. Meanwhile Immanuel claimed that his experiments with "reinforced blasting powder" were proving successful, and the elder Nobel called for Alfred to leave St. Petersburg and return to Sweden to help him with the experiments. Ultimately the concept proved fruitless, because as the nitroglycerin-soaked gunpowder aged, it became weaker and ended up being only marginally stronger than regular gunpowder, but much more expensive. An official trial and test of the fortified powder for the Swedish military was an embarrassing failure, and Alfred later wrote with resentment that he had "wasted the whole summer with experiments which a competent person could have carried out in a day." He quietly returned to his own project.

In the tiny Heleneborg laboratory, Alfred continued to tinker with his design, eventually deciding to place the small charge of gunpowder in the nitroglycerin instead of the inverse, which he had previously done. When it came time for a public demonstration, he proudly brought out his unwieldy contraption while his father and two of his brothers, Oscar-Emil and Robert, looked on. Lighting the fuse, he carefully flung the bomb away and awaited the blast. None came. His father and Robert burst out laughing, and Alfred, humiliated and hurt by their mockery, collected his materials and retreated shamefacedly to the laboratory. He never forgave his father for laughing at him, reporting ten years later during a trial for a U.S. patent: "I had many failures, until my father and brother who witnessed them ridiculed my tenacity."

Why had the device worked when thrown into water but not on land? After mulling over the problem, he realized that the water had served to confine the explosion, thereby generating the necessary force to detonate the nitroglycerin, but in the open, too much of the pressure from the exploding gunpowder escaped and

Alfred Nobel, ca. 1880. He is the inventor of dynamite and other specialized high explosives that transformed Western society during the late nineteenth century. Nobel presided over a mighty industrial empire and accumulated a vast fortune that he left to fund the Nobel Prizes when he died in 1896.

the nitroglycerin failed to detonate. He tested it again and again after sealing the ends of the tube with wax, and was successful every time. He had discovered a reliable means of detonating nitroglycerin. "The real era of nitroglycerin," he later wrote, "opened with the year 1864 when a charge of pure nitroglycerin was first set off by means of a minute charge of gunpowder." It was the Initial Ignition Principle, and it irrevocably redirected the practical and theoretical inquiry into explosives forever. Nobel later settled on a metal tube with a wooden or metal cap of gunpowder suspended in it (and later replaced the gunpowder with mercury fulminate, or detonating mercury). Experiments with mercury fulminate were conducted by a Scottish reverend named Alexander John Forsyth in the early years of the nineteenth century. Forsyth was an avid bird hunter, and he sought to devise a means of igniting the gunpowder in his gun without forewarning his wary quarry with the flash from the pan of his flintlock. He invented the percussion lock—a trigger that released a hammer that would smack a cap of mercury fulminate, setting it off and firing the gunpowder. He wrote in his patent application that "instead of giving fire to the charge in the gun by a lighted match, or by flint and steel, or by any other matter in the state of actual combustion . . . I do make use of some or one of those chemical compounds which are so easily inflammable as to be capable of taking fire and exploding without any actual fire being applied thereto, and merely by a blow." Nobel's process for igniting nitroglycerin was based on a similar principle. Instead of a percussion lock, he used a fuse to supply the initial shock that set off the mercury fulminate, which in turn set off the nitroglycerin (which required a much greater percussive force to detonate than gunpowder).

Alfred rushed to file his invention, which he called Nobel's Patent Detonator, with the Swedish patent office on October 14, 1863, and quickly followed it up with patents for a series of minor improvements throughout Europe and in the United States. "I therefore lay claim to the idea," he wrote, "so far as industrial use is concerned, of contriving by administering a mere initial impulse to develop an explosion in substances which, exposed, can be brought into contact with burning bodies without exploding." He was referring to the curious fact that nitroglycerin would not explode when exposed to flames. It merely burned away like oil, leading Nobel to a false belief in the dangerous substance's safety. During the manufacturing of the oil, however, when the sulfuric and nitric acids were mixed with glycerol, it was particularly susceptible to overheating and at great risk of exploding. Throughout the nineteenth century, a manufacturing process evolved whereby the operator sat on a one-legged chair so he would never get too comfortable and allow his mind to stray from scrutinizing the large thermometer prominently positioned on the mixing vat. This sort of safety precaution, however, was years away from the primitive techniques and equipment in Nobel's early laboratories.

After several well-attended flawless demonstrations at Swedish mines and railway construction camps, the word was rapidly spreading of the potential of this new invention. Nobel made a quick trip to Paris and secured some capital investment from the Pereira bank, which was financing Napoleon III's construction projects, including the Suez Canal, and understandably had a great interest in the prospect of a new and more potent explosive that would speed construction and reduce costs. Nobel had just returned with the money, hired a handyman and a chemist, and indeed was manufacturing the first significant quantities of nitroglycerin for shipment to the Swedish railroad and mining crews when the explosion ripped through the laboratory on September 4, 1864, killing his younger brother and four others.

The day after the explosion Immanuel was summoned to the police station to give a report to the investigating officer, but Alfred presented himself in his place because his father was stricken with grief and was temporarily paralyzed (a month after the explosion Immanuel suffered a stroke and was bedridden for the remaining eight years of his life). The police inquiry was to determine the cause of the explosion and to decide if homicide charges were warranted. Five people were dead, the result of an explosion in a laboratory where nitroglycerin was being produced within the city limits. Alfred and his father prepared their argument, defending their actions by claiming that "nitroglycerin is harmless, even if lighted, and the greatest carelessness in the use of fire is hardly able to produce an explosion. The only possible explanation that remains is that the experiment made by my son produced a violent reaction which raised the temperature of the mixture to the heat of approximately 180 degrees centigrade, at which fully formed nitroglycerin would explode.... Unless heated quickly to 180 centigrade and strongly contained, the nitroglycerin would not explode but burn away." It was Immanuel who held the license for experimenting with minute quantities of explosive material, and he was supposed to notify the local police if he ever began to manufacture dangerous quantities—something he had failed to do. The only excuse the Nobels could offer was that although they were preparing orders for shipment, they weren't manufacturing per se, but were still perfecting the procedure and the final product, before shipping. Some Nobel biographers have suggested that it was the influence of the State Railway and mining interests, desperate for the new explosive, who tilted the police away from charging the Nobels with murder. In the end there were fines to be paid and warnings and a strict policy prohibiting any experimentation and manufacture of nitroglycerin within the Stockholm city limits, including Heleneborg.

Ironically, despite the devastating personal loss and the frightening shadow of homicide charges, the terrifying explosion confirmed the true power of the new explosive, and Nobel's prospective business partner, Smitt, was forthcoming with his offer of investment. By November 1864, a company was formed with Alfred and his father having just under half the shares. Alfred pushed ahead with his research despite the death of his younger brother, his father's stroke, and his older brother Robert's advice "to quit as soon as possible the damned career of an inventor, which merely brings disaster in its train." Nobel's first production factory, a temporary facility to serve until a plant could be constructed in remote Winterviken, was a covered barge anchored near the shore of Lake Malaren outside city limits. He quickly began fulfilling contracts to the State Railway for blasting tunnels. News of the dreadful explosion had not faded in people's minds, however, and he had to contend with the hostility of the locals, who protested and called his mobile factory a "death ship." He was constantly on the move, floating his laboratory to new anchorages, occasionally fleeing when mobs of angry peasants chanted on the shore brandishing pitchforks. Eventually he anchored the barge in the middle of the lake, laboring alone and bundled against the cold, until March 1865, when the facility at Winterviken was operational. Business picked up rapidly from then on, the beginnings of the world's first multinational industrial explosives and armaments empire.

By providing a reliable and relatively safe method of detonation of nitroglycerin, Nobel had given the world an explosive of tremendous power. The chemist and historian G. I. Brown observes that "when nitroglycerine explodes it produces a large volume of hot gasses and a consequent big increase in pressure, just as gunpowder does, but the effect is on an altogether different scale. . . . It is the difference between being bumped into by a pedal cyclist or being knocked for six by an express train." The slower, steady push of gunpowder was suitable for both demolition and propulsion in guns and cannons. The lightninglike crack of nitroglycerin, although too powerful and quick for guns and cannons—it shattered their barrels as often as it launched a projectile—was ideally suited for mining, tunneling, demolition, and excavating. It quickly became indispensable. By the 1860s industrialization was creating an ever greater demand for coal and metals, and the need for better means of mining them. The new explosive in turn inspired and enabled new construction projects, which in turn required more coal and metal. Nitroglycerin, marketed and branded as Nobel's Blasting Oil, rapidly turned around the Nobel family fortune, becoming indispensable and sought-after throughout the world.

By the time Alfred had departed St. Petersburg for good in 1863 and rejoined his father and mother and younger brother, Oscar-Emil, in Stockholm, he displayed signs of the peculiar blend of characteristics that would see him rise to the pinnacle of wealth and power, propelling him on a lifelong quest to devise new and innovative products, yet shrinking and containing his personal world, leaving him lonely and aloof. As he matured, certain traits crystallized: He was enigmatic, introspective, and isolated, with an awkward and morbid sense of humor. He joked about constructing a luxurious suicide emporium in Paris, where people could go to die in plush surroundings while an orchestra played "the most beautiful music." He was also prone to tantrums when he didn't get his way. On one occasion, when he missed a ferry by mere seconds, he angrily threw himself into the river fully clothed and swam across, emerging shivering and disheveled. And he mused darkly about his own life, cryptically suggesting, when asked by a brother to supply a brief autobiography, that "Alfred Nobel's miserable existence should have been terminated at birth by a humane doctor as he drew his first howling breath." But he was also pragmatic, stubborn, brilliantly creative, and driven to a fault. For his fledgling nitroglycerin venture in Sweden, he worked round the clock, serving as managing director, engineer, chemist, traveling salesman, public relations correspondent, treasurer, and advertising manager. Yet out of his first profits he forwarded a handsome sum to his mother so she could afford medical treatment for his recovering and partly paralyzed father. The same traits that secured his triumph as an inventor and industrialist left him little time to pursue other interests and contributed to what he believed were his personal failures, his lack of family and friends. Never in robust health, he was a slight man, with receding hairline, bulging eyes, bulbous forehead, large ears, and a curious crablike retreating manner. He routinely entered his own laboratory by the rear entrance and referred to himself as elderly when he was in his midforties. But nothing in his portraits reflects what would become his defining characteristics—his intense concentration, stubbornness, prideful ambition, and all-consuming working hours. When a problem presented itself he threw himself wholly into the fray, ceaselessly toiling until a solution presented itself. The problems with nitroglycerin began to materialize soon after Nobel branched out from local distribution and expanded his markets.

International demand for his blasting oil proved to be as strong as the optimistic Nobel had predicted, and desperately hoped. Instinctively knowing how easy it would be for others to illegally manufacture his blasting oil and equip it with a version of his patent detonator, and having an appreciation for the difficulties of transporting liquid nitroglycerin—its sensitivity to impact—Nobel began applying for patents throughout Europe. He wanted the substance to be manufactured as close as possible to where it would be used. With separate patents already granted for Finland, Norway, and England, and many others pending, Nobel accepted an offer of partnership from a German businessman in 1866 for a large factory in the suitably remote valley at Krümmel, south of Hamburg, on the Elbe River. From Hamburg, his blasting oil could be shipped anywhere, even as far as Australia. Initially, fifty employees started work at the factory, and the staff soon increased dramatically as the orders poured in from Europe and the world.

Nitroglycerin was packaged in zinc canisters packed in wooden crates stuffed with sawdust to damp potentially destructive impacts. Nobel went on a whirlwind marketing tour of Europe, hauling padded suitcases filled with demijohns of blasting oil to the mining districts for demonstrations, further stimulating interest in the fantastic explosive. Over and over again he demonstrated, and proved, that blasting oil was an explosive of an entirely superior caliber to gunpowder. The evidence was plain to see, and despite its relatively steep price, the orders kept rolling in.

In the United States nitroglycerin was used to blast tunnels through the Sierra Nevada for the Central Pacific Railroad, saving the company millions of dollars and shaving months off the time originally estimated to complete major tunnels. The historian and Nobel biographer Eric Bergengren wrote that "Nobel's discovery that nitroglycerin could be detonated was worth millions of dollars to that one corporation." In Australia it was used in mining and quarrying. In England, the greatest demand was in the slate quarries of North Wales. In Norway and Sweden it was eagerly embraced by miners and the railways. Blasting oil was also being studied for its capabilities in war by military planners in many countries. By the end of 1865 the nitroglycerin business was booming, stimulating a corresponding increase in the demand for nitrates.

Because of nitroglycerin's seemingly harmless and stable nature, people failed to treat it with the caution and respect with which they would have treated gunpowder. Accidents happened. Unbeknown to Nobel or any of his partners, as nitroglycerin aged, the impurities left in it in the manufacturing process caused it to deteriorate and become even more volatile. The acid corroded the seams on the zinc canisters in which it was contained, allowing the blasting oil to trickle out and pool in the holds of ships, where it was buffeted and sloshed, or to ooze out of carts and railcars onto the axles. In one instance, after a demonstration at a quarry in Sweden, the engineer strapped two bottles of nitroglycerin to the roof of a stagecoach filled with passengers, which happily trundled over the rutted, bumpy back roads from a proposed railroad site. The blasting oil, according to the engineer, "oozed over the side of the carriage and the wheels." A bottle fell off the coach and was found by a woodcutter who used it to oil his boots and his horse's harness. At a mine in Silesia blasting oil was kept outside in winter, where it froze into lumps. The mine manager merrily chipped away with his ax until he had the desired quantity. On one occasion in a warehouse in California, leaking canisters were patched by a local plumber. On another occasion, a shipper tried to salvage the expensive oil that was leaking from corroded canisters by transferring it to new tins. Afterward he decided to burn the old rusty tins and they erupted, fortunately not killing anyone.

Other accidents were not so harmless or amusing. In Wales, a miner and his mates were kicking a canister as a makeshift football when it exploded, killing one of them. In November 1865, a traveling salesman from Germany named Theodore Luhrs misplaced a suitcase of blasting oil in the Hotel Wyoming in Greenwich Village in New York City. Several days later, patrons in the barroom complained of an acid stench that was coming from the storeroom. A porter opened a closet and saw the suitcase seeping vapor. He hastily dragged it into the street and ran back inside the hotel just as it detonated, blowing a four-foot hole in the pavement, shattering windowpanes along the entire street, and riddling eighteen people with glass shrapnel. On December 11 of the same year at Bremerhaven, on the north coast of Germany west of Hamburg, a terrifying blast destroyed the steamship *Mosel*, killing twenty-eight and injuring nearly two hundred. Before he died, a fatally injured American named William King Thompson confessed to placing a nitroglycerin bomb in the hold to destroy the ship and earn him a good return on some heavily insured cargo. The bomb went off before he could flee the death ship.

The blast at Bremerhaven was followed in 1866 by a series of explosions around the world—from nearly everywhere Nobel's blasting oil had been shipped. An aging supply of blasting oil in a warehouse in Sydney, Australia, detonated and tore through a warehouse complex, killing twelve. The steamship *European* with blasting oil in the hold spontaneously erupted in Panama, killing sixty and severely damaging the pier and docks. And in April that year, a thunderous explosion in the Wells, Fargo and Company compound in San Francisco demolished most of a block, killing ten and injuring dozens. The *San Francisco Chronicle* from April 21 recorded the community's horror with the headline "Terrible Explosion and Loss of Lives in San Francisco." "A box containing this liquid [blasting oil] had arrived by steamer from the East, and when landed upon the wharf was found to be in a leaking condition. It had been shipped as general merchandise, and none were aware of the dangerous contents of the box." The explosion occurred when two employees tried to open the crate with a crowbar to see what was leaking inside. "The explosion was so powerful as to shake the earth like an earthquake for a circuit of a quarter of a mile. Every window in California Street, between Montgomery and Kearney, was demolished, and panes of glass were shattered over as far as Third Street, a distance of half a mile." The article then contained the following disclaimer and notice:

WARNING!
THE FOLLOWING DESCRIPTION OF THE RESULTS OF
THE EXPLOSION IS EXTREMELY GRAPHIC.

Fragments of human remains were found scattered in many places. In the auction room of Cobb and Sinton, on the east side of Montgomery Street, a human brain, almost intact, and other fragments of the body near it, were found. A piece of human vertebrae was blown over the buildings on the east side of Montgomery Street, where it was picked up in front of Squarza's, on Leidsdorff street. A piece of skull was lying on California Street, east of Leidsdorff, with other fragments of human remains, and a human arm struck the third story window of the building across the street.

In May, while Nobel was in New York seeing to the licensing of his American patent, carrying with him a large padded suitcase stuffed with nitroglycerin for demonstrations, his factory in Krümmel was demolished by an explosion, and a month later the plant at Lysakar in Norway went up. It was a staggering series of blows to his business. Were all these deaths and millions of dollars in property damage to be considered an accident? Who was responsible? The name Nobel was reviled and his operations were shut down in many jurisdictions. A hastily enacted California law prohibited the transportation of nitroglycerin, and other similar laws around the world followed. In France and Belgium, possession of nitroglycerin was outlawed; in Sweden, its transportation was banned; in Britain, laws requiring special permits were enacted, effectively banning the substance. The United States considered making any deaths directly linked to the transportation of Nobel's sinister oil a crime of first degree murder punishable by hanging. The name nitroglycerin sent a shiver up the spine, and tales of its exaggerated power were widespread, inspiring terror and fear. Railroads refused to carry it, ships refused it as cargo, and longshoremen refused to handle it. Even potential customers who would benefit greatly from nitroglycerin's blasting capacity were now frightened off from business dealings with Nobel and his band of devils. But a great deal of blasting oil was being used in the Prussian-Austrian War of 1866.

The article in the *San Francisco Chronicle* summed up the public's sentiment toward nitroglycerin and blasting oil, not only in America but throughout Europe, with its concluding pronouncement: "This must be stopped at once. . . . Public safety demands it." Nitroglycerin was too powerful to be entirely prohibited forever, but the restrictive laws might easily bankrupt the first company to commercially exploit it, a fact not lost on Nobel. Drastic action was needed to salvage the remnants of his neophyte enterprise.

Construction and Destruction

Dynamite and the Engineering Revolution

We tested what we considered a very small quantity, but this produced such terrific and unexpected results that we became alarmed, the fact dawning upon us that we had a very large white elephant in our possession. At 6:00 A.M. I put the explosive into a sarsaparilla bottle, tied a string to it, wrapped it in paper and let it gently down into the sewer, corner of State and Washington Streets.

—*Thomas Alva Edison, Boston, 1868*

Metals have been sought-after substances since antiquity for weapons, armor, tools, and ornaments. Men have scoured the surface of the earth for outcroppings of metal-bearing ore, and when the easily obtained surface sources had been depleted, miners followed the valuable veins into the bowels of the earth. Early mining practices churned through human laborers like a bonfire devouring wood. Conditions were inhumane. In the second century B.C. the Greek geographer Agatharchides toured several Egyptian gold mines and related his horror at what he beheld once past the opening of the mine. Great multitudes of fettered human flotsam toiled away in the darkness, "notorious criminals, captives taken in war, persons sometimes falsely accused, or against whom the King is incensed; and not only they themselves, but sometimes all their kindred and relations together with them."

These miserable masses raced to stack great racks of timber against the rock face and then set the wood on fire. The roaring flames consumed the timber, sucking the oxygen from the pit and emitting clouds of creeping smoke. Overseers ordered brigades of disheveled wretches into the noxious fumes to heave buckets of cold water against the stone in the hope that the rapid temperature change would crack the stone face. Their throats choked on the sulfurous or arsenious gasses, their lungs burned from the cinders of smoke, and their skin was scorched by the steam. "The rocks thus softened and made more pliant and yielding," Agatharchides related, "several thousands of profligate wretches break in pieces with hammers and pickaxes . . . cleave the marble-shining rock by mere force and strength . . . following the bright shining vein of the mine." Little boys "penetrate through the galleries into the cavities and with great labour and toil gather up the lumps and pieces hewed out of the rock as they are cast upon the ground, and carry them forth and lay them on the bank." The slave miners, pasty troglodytes in their subterranean tomb, perished by the thousands, worked to death—crushed by falling rock, suffocated by the fumes of the fires, asphyxiated by pockets of gas, or drowned when their hammering unexpectedly burst unsuspected aquifers.

Agatharchides was sickened and appalled. "No care at all is taken of the bodies of these poor creatures," he related, "so that they have not a rag so much as to cover their nakedness, and no man who sees them can choose but commiserate their sad and deplorable condition. For though they are sick, maimed, or lame, no rest nor intermission in the least is allowed them . . . but all are driven to their work with blows and cudgeling, till at length, overborne with the intolerable weight of their misery, they drop down dead in the midst of their insufferable labours." Ancient Egyptian mining practices reflected perhaps the worst human abuses for centuries to come. Even the later Greek and Roman mines, which also used slave labour, were managed with a greater degree of humanity. But the sorrowful lament expressed by Agatharchides held true for centuries. "I cannot but conclude," he wrote, "that nature itself teaches us, that as gold is got with labour and toil, so it is kept with difficulty. It creates everywhere the greatest cares; and the use of it is mixed both with pleasure and sorrow."

Oddly, the conditions of those who toiled beneath the earth were worse in the celebrated and gilded civilizations of antiquity than during the so-called Dark Ages that followed. Although mining has always been a dangerous occupation, it achieved a sort of golden age during the later Middle Ages, the era of the free Saxon miners in Germany, England, Hungary, Austria, and Bohemia. The German physician Georgius Agricola described a very different scenario from the horrors of Agatharchides' day in his famous mining tract *De re metallica*, a compendium of all contemporary mining techniques then in use throughout continental Europe, published in 1556, a year after his death. At its pinnacle, the Saxon miner was a respected member of a skilled and highly regulated trade, enjoying reasonable work shifts, clean working conditions, and one or two days off per week. "They lighten their long and arduous labours by singing," he wrote, "which is neither wholly untrained nor unpleasing." The use of fire mining was strictly regulated by a regional mining authority that reported to the local lord, and consequently was quite rare. The reason for its limited use, as explained by Agricola, was simply because it was dangerous to the miners. "The heated veins give forth a foetid vapour and the shaft and tunnel are emitting fumes so that no workmen can go down into the mine lest the stench affect their health or actually kill them." When fire mining was allowed, it was with the consent of the surrounding mines and was lit after the last shift on Friday so the men did not return underground until the following Monday.

Despite professional standing and dignity, however, the work of a miner was still brutal. Cold mining of hard rock with nothing other than picks and wedges was backbreaking, physically demanding work, done in stale air with flickering candles as the only light source. Often alone at small mine faces, the miners pounded spikes and wedges into cracks with a sledgehammer, heaved and pried with crowbars to dislodge great chunks of ore and rock, and hauled it slowly to the surface in heavy leather shoulder bags. Progress was slow and erratic, occasionally speeding ahead if a great boulder snapped free, or if fire was used to loosen the stone face, sluggish and frustrating when a hard, smooth surface presented itself.

Around the same time, and particularly by the seventeenth century, the conditions in Scandinavia, particularly in Sweden, were in stark contrast to the conditions for miners elsewhere in Europe. Here fire mining was a standard practice, and the mine churls were expendable bodies, ordered into the fume-filled pits to douse the flames and soak the rock, diving to the earth and covering their faces with wet rags in a vain effort to avoid the burning, poisonous clouds that rolled down the tunnels. The great Falu mine in Sweden burned through over seventy thousand cords of wood annually, soon denuding the countryside of timber to feed the monstrous fires in the pits. The famous Swedish naturalist Carolus Linnaeus visited the Falu copper mine in 1734 and recorded what he saw. It was an experience that made him shudder, for these were not slaves and war prisoners but his fellow countrymen who toiled under barbaric and inhumane conditions. "The mine is more terrifying than the Hades as described by classical writers and Hell as painted in the sermons of our clergy," he wrote. "The sulphurous smoke poisons the air and kills everything growing, and fills the cavities of the mine with evil fumes, dust and heat. Here 1,200 men labour shut off from the light of the sun, slaves under the metal, less men than beasts, surrounded by soot and darkness. . . . These *damnati* work naked to the waist and have before their mouths a woolen cloth to prevent them from breathing too much smoke and dust. They cannot take a breath of pure air, sweat streams from their bodies as water out of a bag." Scandinavia at the time was one of the great sources of valuable minerals that were a prime source of wealth for its rulers, and so the work went remorselessly on. According to the historian Gösta E. Sandstrom in *The History of Tunnelling*, mining "as practiced in the Falu mine, constituted a singularly outstanding example of brutal and technically backward mining practice." But many other mines throughout Europe soon turned to similar practice. The demand for metal ores had outstripped the capacity of the hand-powered, muscle-driven cold mining methods of the Golden Age of the medieval miner. Fire mining, though barbaric, cruel, and destructive, and discounting the terrible cost in human suffering and death, was a far more efficient method of extracting ore. By the early seventeenth century, the requirement for greater productivity had eclipsed the centuries-long tradition of humanity, respect, and compassion for one's fellow man.

Since the dawn of humanity, energy has been harnessed by muscle, fire, wind, and water. In the thirteenth century gunpowder was added to humanity's list of powerful tools and was soon in common use by soldiers. Sometime in the mid-seventeenth century gunpowder began to be used in mines throughout Europe, probably due to the experience of discharged soldiers returning from wars who had used it to undermine fortifications. Historians from various countries, such as Germany, Austria, Hungary, and Sweden, have argued for primacy in the first use of gunpowder in mining. "There is always a desire to pin down the breakthrough to one particular person at one particular time," notes Sandstrom. "In their patriotic ardour the historians of different countries are anxious to prove that one of their own compatriots was the inventor, and they waste much effort heaping up proofs to this effect." Determining the first use of gunpowder in mining is not only nearly impossible, but also irrelevant, as it appeared around the same time throughout Europe. The rise in gunpowder use for mining and other civil engineering projects coincided with the rise of the saltpeter trade with India in the mid-seventeenth century, which reduced the cost of saltpeter, and hence the cost of gunpowder, as well as freeing up the limited supply. When saltpeter became more readily available as an import commodity, rulers who had previously hoarded the substance for military use relinquished excess supplies for the task of liberating ore from the earth and for blasting routes for canals and roads. By the eighteenth century, gunpowder was an indispensable and standard tool for mining throughout Europe, displacing fire mining in the vast majority of mines.

Early gunpowder mining called for a fearless, or likely desperate or starving, miner to crawl to the farthest crevasses of the mine with a bag of the volatile powder and stuff it or pour it into cracks in the rock. Then he laid out a long straw or a series of goose quills filled with powder along the ground. After lighting the fuse he ran, occasionally not fast enough, to escape the terrific explosion that pulverized great slabs of auriferous stone from the mine face into the tunnel. Later techniques involved drilling holes into the stone and then plugging the powder hole with a wooden or clay cap to seal the explosion and increase the pressure. It was exceptionally dangerous work, particularly as the only light came from burning candles or oil lamps. Many men were blasted to smithereens when a stray spark ignited the powder, bringing down the cavern on top of them. A contemporary scene of a mishap in the Falu mine shows the workers cringing, hands flung in the air in a vain attempt to ward off flying chunks of stone as the mine face violently collapses, toppling ladders and scaffolding. Nevertheless, with gunpowder blasting productivity surged. As the diggings progressed ever deeper into the earth, other hazards beset the hapless miners, such as a lack of oxygen and the presence of naturally occurring gasses, such as methane, that could be ignited by a spark from a metal tool or by the open flame of the lanterns. Occasionally the dampness in the underground workings made it difficult to get the powder or fuse to ignite properly, and men were killed in belated explosions as they crept up to the mine face to investigate unexploded gunpowder. The incidents of unplanned explosions diminished somewhat after 1816 when Sir Humphry Davy developed a safety lamp, and further after 1831 when William Bickford invented the safety fuse, a textile-wrapped cord with a black powder core that provided reasonably safe, accurate, and timely detonations.

Accidental explosions were so common that the handling of gunpowder took on a reverential air. Late-eighteenth-century instructions for the use of blasting powder at the Royal Gunpowder Factory at Waltham Abbey in southeast England are amusing by modern standards but reflect the respect and caution that was demanded of early powder workers. "Whosoever is at Labour within or without the powder magazines should execute his commission in such a respectful and revered silence as is seemly in such a place where (unless the Almighty in his Grace keeps a protective hand over the Labour) the least lack of care may not alone cause the loss of life of all present, but may even in a moment transform this place as well as its surroundings into a heap of stone . . . be it out of annoyance at Labour or still less out of lack of faith [men] are most earnestly beseeched not to let emane from their mouths oaths or swearwords or other light or obscene language, whereby the name of the Lord is dishonoured and taken in vain."

The use of gunpowder for civil projects slowly changed the face of the European countryside. It enabled previously inconceivable nonmining and nonmilitary projects such as blasting great troughs for canals, leveling hilly ground for rail beds, and the first attempts at subaqueous tunnels. The first great engineering project to be undertaken with gunpowder was the Languedoc

Canal, or Canal du Midi, carved through the southwest portion of France, linking the Mediterranean Sea and the Bay of Biscay in 1681. At over 148 miles in length, with water levels regulated by 119 locks, and including a 165-meter tunnel, peaking at over 620 feet above sea level, it was a phenomenal feat of daring-do and technical prowess rivaling the ancient aqueducts of the Romans over a millennium earlier. Patrick Beaver, writing in *The History of Tunnels,* observed that "it must have been a trying and violent task. The problems involved in the use of explosives in the confined space of a tunnel and the behaviour of rock under such conditions had never been studied and it is unlikely that the work was carried out without considerable loss of life." Despite the logistical challenges and inherent dangers, canal building continued at an escalating pace into the nineteenth century. Canals were slowly and painstakingly carved across the European and eastern North American countryside. Between 1817 and 1825, for example, gunpowder was used to blast the 585-mile-long Erie Canal, providing a waterway for commerce and travel from Buffalo, on the shore of Lake Erie, to Albany, on the Hudson River, and thence, via the river, to New York. Two years later construction was completed on the Welland Canal, which links Lake Ontario with Lake Erie and bypasses Niagara Falls, extending oceangoing transport to the Great Lakes. Fresh fruits and vegetables could now travel quickly to key urban centers, and bulky commodities like coal could more easily be transported from the mining regions to burgeoning industrial centers. By the mid-nineteenth century, there were over 7,000 kilometers of canals in England and 75 kilometers of canal tunnels, linking the entire southern portion of the island in a web of interconnected river transportation routes.

Despite the vast quantities of gunpowder being manufactured and used and the steady improvements in its blasting strength as the blend of the ingredients was perfected, as the nineteenth century progressed the wild dreams of civil engineers had again outdistanced the explosive capacity of gunpowder. Projects had to be placed on hold, such as blasting in or under water, and building long tunnels through the hard rock of the Alps and the Rockies that would enable year-round transportation and communication. Because of its bulk, gunpowder could not be put in small cracks in sufficient quantities to efficiently break hard rock. One example from the 1860s reveals just how absurd the situation was getting with the great mountains of gunpowder being heaped up to try to blast tunnels. One of Nobel's business partners, Otto Burstenbinder, described the awkward and inefficient predicament faced by the contractors of the Central Pacific Railroad in the Sierra Nevada. "Have you ever seen mines in which three hundred to seven hundred kegs of gunpowder are exploded at once?" he wrote. "We have such mines. . . . A tunnel is made into which seven hundred kegs, with twenty-five pounds of gunpowder per the keg, are placed and exploded. This method, however, costs much time and money. . . . The Central Pacific Railroad alone consumes about three hundred kegs of gunpowder a day." The quantities of gunpowder needed for breaking hard rock were too large to be feasible in a rapidly industrializing society. Western civilization was on the verge of a golden age of engineering, tethered only by the limited explosive capacity of gunpowder, which was just too weak to accomplish what entrepreneurs and engineers now knew to be possible, and, according to the thinking of the day, believed to be desirable.

When Nobel revealed the power of nitroglycerin, it seemed almost too good to be true—it was the solution to a great many of the problems then being faced, an enabling agent for the aspirations and dreams of a wildly optimistic and grandiose age. At nearly seven times the strength of gunpowder, small quantities could easily be placed in boreholes or cracks in rock, astronomically improving the rate of blasting for tunnels and mining. Then came the series of terrible explosions that shattered the foundations of Nobel's emerging enterprise and sent crashing the soaring dreams of the promoters and entrepreneurs. Nitroglycerin was just too dangerous. It had to be tamed.

When Alfred Nobel sailed back to Germany in August 1866, departing America in disgust over patent and legal shenanigans, never to return, his business empire was foundering in the face of an international public outcry against his infernal product. Instead of giving up in resignation, the tireless thirty-three-year-old business operative immediately holed up in a makeshift laboratory near the ashes of the Krümmel factory and continued his life-threatening experiments to stabilize nitroglycerin. "As early as 1863," he wrote, "I was fully aware of the disadvantages of nitroglycerin in its fluid form," and he "saw the disadvantage of a fluid explosive and set about finding a means of counteracting this drawback." Although his initial attempts to improve the safety of nitroglycerin centered on reducing its volatility as a liquid, he strove now to find a means of making it into a solid, to avoid leakage and make it easier to use in blasting. He tried unsuccessfully to mix nitroglycerin with a wide array of substances, including powdered charcoal, sawdust, rock powder, paper pulp, brick dust, coal dust, and gypsum, among others. According to one story, it was an accident that led Nobel to the perfect substance. At the Krümmel factory, instead of the usual sawdust used for stuffing crates of nitroglycerin, a type of widely available local clay named kieselguhr was used. When a worker carelessly dropped a box of nitroglycerin canisters, the deadly oil oozed over the clay and Nobel noticed that it had been completely absorbed, forming a granular puttylike mass. And so, the story goes, one of the world's greatest discoveries came about by chance. Nobel persistently denied the rumor throughout his life, a claim corroborated by his assistants and his own correspondence. According to

Nobel, it was after months of testing dozens, perhaps hundreds of possibilities, that he settled on kieselguhr.

Kieselguhr is a porous, inert, naturally occurring clay, and Nobel determined that a blend of 75 percent nitroglycerin to 25 percent cleansed clay would produce the most stable explosive. He patented it under two names: Dynamite, after the Greek *dynamos* for "strength," and Nobel's Safety Powder, to conjure up and promote an image of an innocuous and harmless product. The malleable putty was shaped into sticks and coated with a stiff paper cartridge, ready-made to fit into standard mining boreholes. Although it was weaker than pure nitroglycerin, it was still five times as powerful as gunpowder. It came in an ideal shape and was easy to handle, to detonate, to store, and to transport. Nobel's business prospects were again on the rise.

Dynamite ushered in a new era of industry and mining. It caused a profound sea change in the way civil engineering projects could be conceived and executed. The age of explosives had arrived. Anything gunpowder could do, dynamite did far more safely, far more efficiently, and far more cheaply. Dynamite rapidly became the most sought after explosive in the world, and as soon as it was in production in 1867, Nobel stopped shipping nitroglycerin entirely (particularly after another blast in a railway car in Belgium killed eleven people). Nobel worked incessantly; he had no life outside of his laboratory and no time for pursuits other than business and explosives research. He personally oversaw the creation and management of nearly a dozen new factories in a dozen countries, but he was frequently ill because of his natural weakness and overwork. According to Eric Bergengren, one of Nobel's biographers, "During the ten years following the discovery of dynamite (1867–1877), when factories and their organization were expanding, Alfred Nobel lived a restless and nerve-racking life which was dictated by the force of circumstances, and he spent much time travelling. These years laid the foundation of his success and his fortune, but they also cost him worry and toil which bordered on the incredible. They left their mark on his health, and on his view of people and the world around him." The results of his incessant labor were impressive. Dynamite had become so popular as the preferred explosive for nearly every blasting use that it spawned countless imitators and soon became a generic term for all manner of explosives that were based on stabilized nitroglycerin. But Nobel's product was still the most reliable, and he soon became a very rich man. In 1867, production of dynamite at his handful of factories was approximately 11 tons. The following year production expanded by a factor of seven to approximately 78 tons, then more than doubled to 185 tons the next year. And continued to double each year, so that by 1874 his factories were producing an astonishing 3,120 tons of dynamite and shipping it worldwide. Within a further decade he had factories in Germany, Finland, Sweden, Norway, Scotland, France, Spain, Switzerland, Italy, Portugal, Austria-Hungary, and several in the United States. Within another ten years there were Nobel, or Nobel-licensed, factories in Australia, Brazil, Canada, Japan, Greece, Venezuela, South Africa, and Russia—a total of 93 explosives factories.

In 1873, Nobel moved from Hamburg to Paris, the intellectual and cultural center of Europe, where he purchased a small mansion in a fashionable district. It was the first home he had that reflected his new wealth and status. Bergengren has noted: "He fitted it out in a manner of dignified, solid comfort, according to the taste of the age, with handsome reception rooms, a wintergarden with hothouses for orchids, in which he was interested, and stables for his fine carriage horses, one of his few hobbies." Although Nobel made an effort to become part of the small Swedish community, and occasionally socialized and entertained, most of his time was still consumed with work. Incredibly, Nobel continued his experiments even while overseeing the expansion of his explosives empire, devising several different varieties of dynamite suited to specific tasks. He called these dynamite 2 and dynamite 3. They were not as powerful and were better suited to coal mining. Other brand names included Atlas Powder and Hercules Powder, and Judson Powder in the United States—within a few years there were countless names and variations on what was essentially the same thing, nitroglycerin combined with an inert substance. Within a few years of dynamite's discovery, chemists and entrepreneurs from around the world began experimenting with methods for improving the power of dynamite by finding a different substance with which to blend the nitroglycerin, one that would not be just dead weight.

Nobel, not surprisingly, was at the forefront of this research as well. His years in Paris were creative and he came upon several important dynamite improvements. He strove to capture the best elements of liquid nitroglycerin, its sheer power, with the best elements of dynamite, its stability, transportability, and safety. In 1875, in his small personal laboratory connected to his Paris home he came upon such a substance. One exhausting but typical twelve-to-fourteen hour workday he was trying to blend, as he had many times in the past, nitroglycerin with a dangerous but powerful explosive substance called nitrocellulose (a blend of concentrated nitric and sulfuric acid with cellulose, an organic substance found in plant tissues). His efforts had failed repeatedly. As Nobel told the story, he accidentally cut his finger and applied collodion (nitrocellulose with a lower percentage of nitration) to form a rubbery protective coating over the cut. The throbbing pain kept him awake all night, and the unfathomable pathways of his wildly inventive and focused mind led him to contemplate that a substance with a lower level of nitration, perhaps similar to the collodion on his finger, might blend well with nitroglycerin. At 4:00 A.M. he rushed down to his laboratory in his nightclothes and began working. When his assistant arrived for work in the morning, he already had the first prototype, a quivering dish of innocuous-looking stiff jelly.

After weeks of work and hundreds of further experiments, testing innumerable substances in different percentages, he settled upon the final recipe. Nobel was not a formally trained chemist; he was a tenacious and creative experimenter. There is no specific record of the exact procedures he followed in his laboratory, the catalysts he tried, or the temperatures with which he experimented. Perhaps he feared the secrets of his research being revealed to his competitors. Certainly it was painstaking, nerve-racking work, locked up in his lab for days on end, breathing the stale air and chemical fumes that gave him headaches so painful that he was forced to lie down on the floor with his head wrapped in a cold compress. He was starved of social contact and exercise and constantly under the strain of time—to solve the problem before others did—and the stress and fear of an accidental explosion.

Nobel called his third great discovery gelignite or blasting gelatin. Blasting gelatin was another milestone that increased the power of dynamite. It was relatively safe, insofar as any explosive is safe, it was easy to mold into the right shape for blasting, it was more powerful than liquid nitroglycerin because both the nitroglycerin and the collodion were explosive, and it was resistant to moisture, making it ideal for blasting under water. But it was too powerful for ordinary uses; it served best in hard-rock blasting for deep mountain tunnels and for the military.

With dynamite and then blasting gelatin, Nobel had revolutionized the way mining, tunneling, and quarrying were done. They opened up great possibilities for industry and civil construction projects. Easy access to coal deposits fueled industrial growth. Cement and concrete became more common construction materials when gypsum and lime could be blasted. Fields could be cleared for agriculture much quicker as a carefully placed stick of dynamite demolished stumps and shattered boulders. Oil

Dynamite was ideally shaped for insertion into bore holes, radically improving the productivity of mining, quarrying, and tunneling.

could be located by setting off dynamite blasts and tracking the vibrations of the earth. Ruined buildings could be demolished in moments. Irrigation and transportation canals could be dug in a fraction of the time they had taken with gunpowder. With dynamite, humans began to rapidly transform the landscape to suit their immediate needs. All of the great hallmarks of nineteenth-century industrial and technical know-how, the famous monumental projects that defined the era, were conceived and completed, with great loss of life, terrible suffering and injury, and in some cases incalculable and irreparable environmental damage, only because of the frightening power of dynamite and its variations. It is for good reason that Nobel is regarded as the father of the modern explosives industry, and although others also invented and improved upon his initial discoveries, and dozens of different brands were on the market, none stands taller than Nobel in truly revolutionizing the industry and, indirectly, the world. The projects that were completed within three decades of dynamite's invention resonate through time as the pinnacle of humanity's efforts to manipulate and change our physical environment. Several decades after the invention of dynamite, the world had been fundamentally and permanently altered.

For centuries a primitive trail wove up through the stunted wind-lashed trees to the rocky saddle of the frequently storm-bound Saint Gotthard Pass at 6,927 feet. Since Roman times the pass was a north-south travel route linking Milan via Zurich to the Rhine Valley. Medieval pilgrims had by the sixth century enlarged the trail to a thirteen-foot-wide pack road for laden beasts en route to Rome, and monks had built a hospice at the summit to shelter wayfarers and give them respite from the storms and cold, as snow lingered late in the high alpine pass. In the middle ages, Emperor Frederick II proclaimed free access to the pass by a charter recognizing its importance. As traffic increased throughout the seventeenth century, the trail was again widened and strengthened into an eighteen-foot-wide road where carriages trundled happily along, and by the early nineteenth century, mail and an ever increasing quantity of goods were hauled over the pass. The road halved the two-week journey between Zurich and Milan. Still, weather shut the pass for much of the year, forcing travelers on long and potentially hazardous detours.

Nineteenth-century railway hysteria overcame even these remote mountain valleys, and in 1853 eight of the nearby cantons agreed to finance a railway tunnel through the pass. But the cost and lack of appropriate blasting technology caused them to abandon the scheme. Ten years later the idea again surfaced, and was again dismissed when no feasible plan presented itself on how to proceed with the daunting and nigh impossible task. After the invention of dynamite, the plan was revised. Eventually it was decided to construct a great bore straight through the middle of the mountain, despite concerns that it might prove impossible. A nearly ten-mile-long tunnel had never been built anywhere. Nevertheless, the Herculean feat was put out to tender to private contractors.

The unfortunate contractor who won the bid was a sanguine, well-groomed, middle-aged man named Louis Favre, a Swiss engineer who rashly agreed to put up the bulk of his considerable fortune to guarantee the completion of the tunnel within eight years. It was a gamble that ultimately cost him his life, his immense fortune, and the lives of hundreds of his men. The piercing of the Saint Gotthard Massif by a tunnel, the greatest construction and engineering feat of the nineteenth century, was to take a decade to complete, cost double the original estimates, and to become a living hell for thousands of ill-treated laborers. In excess of a thousand tons of dynamite were exploded to blast the mighty hole through the mountains between Italy and Switzerland.

In the ill-lit gloom of the bowels of the Saint Gotthard Massif work continued relentlessly twenty-four hours a day, month after month, year after year, as the two openings of the tunnel drove through the solid rock to meet under the mountain. Thunderous explosions shattered tons of rock from the mountain while hundreds of men scampered over the piles of pulverized boulders to clear the way for the next blast. Water gushed from the cracks and boreholes, soaking the men and occasionally requiring them to drag boulders and debris from the tunnel face while wading up to their knees. The temperature rose to a stifling 106 degrees Fahrenheit. When the laborers dropped from heat exhaustion, their comrades rushed to carry them from the confines of the pit. Rock dust, deadly fumes, the stale exhalations of men and pack animals, and the heat brought on a bewildering array of ailments such as silicosis, bronchitis, pneumonia, and miner's anemia, which was caused by a parasitic intestinal worm contracted in the squalid labor camps.

In 1878 *Harper's New Monthly Magazine* recorded the astonished views of an American journalist named S. H. M. Byers, who detoured from his trip to the World Exhibition in Paris to see the celebrated engineering marvel. "As we rushed by dripping walls," he wrote, "and saw here and there ghoul-like figures with dim lamps hiding behind rocks or in deep niches, I involuntarily recalled what our conductor had said of a glimpse of the bowels of hell. . . . The air was so thick that lights could not be seen twenty yards ahead of us, and we all walked close together for fear of being lost or tumbling into some subterranean hole. Far ahead of us we heard the dynamite explosions, sounding like heavy mortars in the midst of battle." If the tunnel workings were a shock to the wide-eyed journalist, the conditions of the workers shocked and appalled him even more so. French, German, and Swiss laborers refused to work in the tunnel under the demoralizing conditions and for the humiliating pay. The bulk of the workforce consisted of Italian peasants. "Their food is extremely limited in quantity," he wrote, "and is wretched in quality, consisting largely of polenta, or a sort of Indian meal porridge. Meat they never taste at all. They are contented to receive their forty or fifty cents a day for hard work, if they can only escape wounds and death from the bad gasses and the thousand accidents to which they are liable every moment of their lives in the tunnel. Alas! They do not escape, for every week records its disaster, either from explosions and flying rocks, falling timbers and masonry, or railway accidents, breaking machinery, etc."

The tunnel eventually claimed two hundred and seventy-seven lives and seriously maimed thousands of others. As many pack animals as men perished in the tunnel. On average one miner died every two weeks, while hundreds of others succumbed to one of the numerous diseases that raged through the work camps of weary, malnourished, and exhausted tunnelers. Several months of labor usually sickened a man, while a year in the diggings left him permanently maimed and ill. One of the dead was Favre himself. As the deadline came and went and the tunnel at least a year from completion, his significant deposit was rapidly being forfeited according to the terms of his foolhardy and punitive contract. He spent days on end in the fume-infested tunnel urging the men on, striving for greater speed, until they literally dropped in their tracks, or made deadly mistakes. Hounded by the railway company and seeing his own personal financial ruin, poisoned by his endless hours in the tunnel overseeing the work, on July 19, 1879, he clutched at his chest and sank to his knees in the muddy ground. He was dead before he could be brought out of the gloom. His corpse was carted out and buried in the cemetery of Göschenen along with the hundreds of other tunnel victims. A year and a half later the tunnel was finally completed and the Saint Gotthard Railway Company offered Favre's now-impoverished daughter a paltry and meager annuity to compensate her for her loss of everything.

With the completion of the tunnel, however, travel time between Lucerne and Milan shrank from more than twenty-seven hours to just over five hours. Today, in addition to the rail tunnel, a vehicle tunnel bores through the Saint Gotthard Massif—the primary transportation corridor for goods shipped between Italy and northern Europe.

On Sunday, September 24, 1876, a great crowd of fashionably dressed men and women gathered along the strand of New York's East River eagerly anticipating a great spectacle. A contemporary sketch shows a respectable crowd congregated along the bank: gentlemen attired in sober suits and hats posed nonchalantly, leaning on their umbrellas, for it was a day punctuated by sudden bursts of rain. Throughout the crowd was a smattering of women with elaborate hats and voluminous dresses, clinging to the arms of their escorts. Elsewhere, perhaps where the view was less clear, other less formally attired folk also pressed for a glimpse of the frightening, yet monumental, event scheduled for that afternoon. And the water was covered with spectators in "a dozen steamers, scores of sailing craft, and myriads of row-boats . . . heavily laden with the endless and innumerable multitude." At least a hundred thousand spectators of all classes of society milled about, but fortunately "great care was taken to keep the liquor shops closed, and but little drunkenness was seen." Despite protests concerning the "unnecessary desecration of the Sabbath," the *New York Times* eagerly reported that "there were never before so many people collected on this continent, perhaps nowhere in the world, to witness any act attempted by human skill." Former lieutenant colonel of engineers John Newton, along with, to allay public anxiety, his wife and two-year-old daughter, was stationed much closer to Hallett's Reef near Hell Gate, a giant submerged outcropping that blocked the channel between the East River and Long Island Sound. The reef created treacherous and swirling currents that damaged about one in fifty of the sailing vessels that dared to push through the waterway and made the channel unnavigable to larger ships.

At the appointed time of 2:30 P.M., Newton, "with all the nonchalance of a blasé disciple of fashion asking some star of the ball-room for a second waltz," guided his toddler's hand to turn a small key, which sent an electric impulse to forty-eight thousand pounds of dynamite and blasting gelatin that had been strategically placed in boreholes and caves around the reef during the preceding years. The explosives detonated with a thunderous dull boom, a furious geyser of water shot into the sky, the earth shook slightly as the submerged reef was shattered into rubble, and "the odors of combusted nitroglycerine swept over unfortunate Yorkville—as though a hundred tallow factories were on fire." The reef was demolished and the *Times* pronounced it a "complete scientific success," although the spectators were disappointed and "considered the whole thing one grand humbug, and if they had paid to see it, a great many would have indignantly demanded that their money should be returned." Although successful, the blast was only the first part in the Herculean task of clearing the channel. Nine years later, on October 10, 1885, Newton repeated his feat when he deployed, for the time, an incredible 283,000 pounds of dynamite and blasting gelatin to finish the job by destroying nine acres of reef known as Flood Rock in Hell Gate, vastly improving the channel for shipping. The *New York Times* reported a day after the detonation: "When those shattered pieces have been gathered up and taken away by the dredgers Hell Gate will have lost its dangers and the wrinkled front of navigation through the Sound will have been smoothed into an inviting smile. Ocean steamers will find twenty-five feet of good, clear water over the once treacherous bottom and a new highway will be open for the commerce of the world."

Another great engineering feat made possible by dynamite was the Severn Tunnel in western England, begun around the same time as the Saint Gotthard Tunnel, to bring a railway under the Severn River to avoid a bridge or steam ferry. At 4.5 miles long, it was the first significant submarine tunnel. Extreme flooding nearly killed dozens of men and caused numerous delays, prompting the tunneler Thomas Walker to increase the working shift from eight hours to ten to make up lost time. Dynamite fared poorly in the tunnel because in the cold temperatures it produced "such deleterious and even dangerous fumes that it was abandoned altogether." Dynamite fumes were frequently a health hazard in poorly ventilated tunnels and mines, although this didn't deter construction efforts. Despite the poisonous fumes and the innumerable flooding incidents, trains began using the Severn

Louis Favre, the unfortunate Genoese mining contractor who lost his fortune, his life, and the lives of 277 of his men while completing the St. Gotthard Tunnel in the 1880s.

Tunnel in 1886, thirteen years after blasting began, significantly reducing the travel time between Bristol and Cardiff.

Other grand rail tunnels of the era include the mile-long Musconetcong Tunnel in Pennsylvania, completed in 1872; the "Great Bore," the five-mile-long Hoosac Tunnel in Massachusetts, completed in 1876 after twenty-one years of blasting, first with gunpowder and then rapidly completed with nitroglycerin and dynamite (two hundred workers died in blasting accidents); and the Simplon Railway Tunnel through the Alps connecting Switzerland with Italy, blasted between 1898 and 1906. Although nearly as long as the Saint Gotthard, at 12.3 miles, Simplon was built with only a fraction of the time and cost and with thirty-two deaths and around eighty-four permanent maimings. Countless hundreds of other lesser rail tunnels were built during this era. In the 1880s, the railway contractor and financier, William Cornelius Van Horne, had three dynamite factories built in northern Ontario that supplied the explosives needed to blast through the Canadian Shield and the dozens of passes and tunnels of the Rocky Mountains and Selkirk Mountains for the Canadian Pacific Railway. Without the power of dynamite and blasting gelatin, a transcanadian railway, considered one of the prerequisites for the western colony of British Columbia joining the eastern part of the country, would have been impossible. Hundreds, if not thousands, of ill-treated Chinese coolies perished in blasting accidents during the construction.

Nobel's dynamite not only built the world's great railways, but also the world's great canals. Between 1881 and 1893, dynamite was used to blast the channel of the Corinth Canal, a 259-foot-deep gorge slicing through the four miles of rock connecting the Peloponnesus to the mainland of Greece. The canal is a steep V-shaped trough carved through rock, twenty-three meters wide, linking the Ionian Sea to the Aegean Sea, and vastly improving the profitability of Athens as a shipping depot. The Suez Canal, a hundred-mile-long Egyptian canal orchestrated by the French engineer Ferdinand de Lesseps between 1859 and 1869, created a commercial waterway between the Mediterranean and the Red Sea, eliminating the circuitous voyage around the southern tip of Africa. It was sped up immensely in the final years by the use of dynamite instead of gunpowder. The 360-mile-long New York Barge Canal connecting the Hudson River and the Great Lakes was completed between 1904 and 1918, with dynamite enlarging and deepening the Erie Canal, which was built with gunpowder eighty years earlier. The canal was critical to the economic development of New York City.

Dynamite also made possible the construction of the Panama Canal, a great, dirty fifty-one-mile-long ditch through the rock across the Isthmus of Panama, the narrow stretch of land between the Atlantic and Pacific Oceans. It liberated ships from the fickle and treacherous voyage around the Horn of South America, shaving weeks off the oceanic shipping time for goods bound between eastern and western North America. The Panama Canal was one of the most expensive engineering feats of the time and it claimed nearly thirty thousand lives before it was completed. De Lesseps, the same French promoter and engineer who had managed the construction of the Suez Canal, began work on the Panama Canal in 1882, but malaria, yellow fever, and cost overruns drove his company bankrupt. The U.S. government acquired its assets in 1904, and continued the colossal undertaking until 1914, when the first ship passed through from the Atlantic to the Pacific. It was the most expensive capital project undertaken by the American government at that time. Millions of tons of dynamite were used to break the earth, and 184 million cubic meters of earth were excavated. Dynamite deepened and widened the Danube River at the Iron Gates in Romania between 1890 and 1896, improving shipping on the mighty eastern European artery. It was used to carve out the world's first underground railway, the London Underground, beginning in the 1860s and expanding throughout the remainder of the nineteenth century. Nearly 10 million tons of dynamite blasted the first tunnels of the New York subway, the first branch of which was opened in Manhattan in 1904. Dynamite also enabled other less pragmatic projects, such as roughing out the figures for the Mount Rushmore sculptures, mighty underground sewers, aquifers for drinking water, and irrigation. Dynamite made possible Australia's great mining boom. Starting in the 1930s, countless millions of pounds of dynamite blasted the canyons and foundations of monstrous hydroelectric dams such as Boulder, Shasta, and Grand Coulee. Industrial-scale mega projects like these would have taken a thousand laborers a thousand years to do what dynamite and its numerous imitators achieved. The number of commercial canals, railroads, and other projects of lesser pedigree created with dynamite is inestimable, and the economic benefit from reliable transportation of bulky merchandise and produce is incalculable. There is no doubt Europe and North America would have evolved very differently without these canals and railways.

These technical marvels, conceived and completed within a span of several decades after the invention of dynamite, completely transformed Europe and America and Australia within a generation. The work was not completed without tremendous cost in human lives and unimaginable suffering. Instead of being content to judiciously use dynamite to more safely and humanely continue with the scope and scale of mining and tunneling projects already enabled by gunpowder, boosters, promoters, and dreamers pushed the envelope of possibility, striving to do what had never been done before. The motivation for the reckless, crazy, and dangerous engineering projects of the dynamite era is the same fundamental human drive that pushes adventurers to summit the world's highest peaks, voyage into the remotest and most dangerous waters, or ride rockets to the moon. A new tool such as dynamite was simply the means to a yet more difficult goal.

Human life in those heady regulation-free days ranked third behind money and time. Vast gangs of illiterate rural laborers paid the highest toll—ill paid, poorly housed, poorly fed, and dangerously exploited, they died by the thousands. These monumental projects extracted a terrible toll on nearly everyone involved, including the head contractors. The historian Sandström noted: "The stresses put on an old-time tunneling contractor, as we have seen, were frequently too much for the human flesh to bear. None of the great tunnellers of the last century reached sixty years of age. Not one of the contractors of the three Alpine tunnels, Frejus, Gotthard, Simplon, lived to see the results of his gigantic efforts. All three dropped dead on the job." Certainly they accomplished more than any workforce in history, breathtaking and truly astonishing feats, but the working conditions did not improve along with their productivity, and their sacrifice is scarcely remembered by the millions of people who routinely travel and benefit from these great engineering marvels. For nearly every one of these projects, the road was littered with casualties. Every mile of the Gotthard Tunnel, as the most extreme example, is stained with the blood of nearly twenty-five men. Until experience had shown the type of dangers to be encountered, and revealed the physical limitations of the human body, ignorance and hubris exacted a terrible toll. Perhaps it had to be this way, despite the horrendous cost, to learn firsthand from bitter experience.

Within a few years, and in nearly every aspect of civil engineering, dynamite modernized the world and led to even more ambitious undertakings and technical experimentation. This included the quest to invent even more powerful explosives—a quest that has not always been motivated by a desire to improve the good of humanity. The primary use of explosives since the invention of gunpowder has been for the military, and gunpowder and dynamite have had at least as great an impact on global history in the field of violent conflict as in civil engineering.

The Great Equalizer
Explosives and Social Change

"The diabolical invention of artillery allows the cowardly villain to cause the death of the bravest of heroes. A bullet comes from who knows where fired by some cowardly wretch who is himself terrified and will run from the flash of his own weapon."
—Miguel Cervantes,
Don Quixote, 1605

In the battle-weary lands of northwestern France, where roving English armies had been pillaging and ravaging the countryside for over a century, the spring of 1449 was about to bring a great change. The ice and snow had departed, and it was now marching season. On April 15th, Sir Thomas Kyriel was leading an English army of about 4,000, including mounted men-at-arms, pikemen, and nearly 2,900 famed and feared longbow archers, to relieve an English garrison besieged by a French army near Caen. English troops garrisoned dozens of fortresses and towns throughout Normandy, tenaciously clinging to lands along the northwest coast that had been won during the Hundred Years War beginning with the Battle of Crécy in 1346, and kept from the French kings by the overwhelming power of the longbow, a weapon of great force that was exceedingly difficult to master but devastating because it could hit enemy cavalry and infantry before they could close for combat.

After several hours of marching along roads through lush green countryside, two riders thundered toward Kyriel's force, reined in, and breathlessly delivered exciting yet disturbing news—a French force of nearly three thousand mounted knights and hundreds of local infantry was deployed along their line of march. Kyriel ordered his men to form the usual defensive formation in a nearby orchard—dismounted men-at-arms were interspersed with clusters of archers on the flanks—and awaited the French army. The Duke of Clermont, commander of the French forces, was well aware of the devastating power of the English archers and kept his men about three hundred meters away, out of range of the deadly iron-tipped shafts. After a few halfhearted probing cavalry advances, which were repulsed by the archers, he brought his new cannons to the fore. He had several large culverins, or long bombards, that were being dragged on horse-drawn gun carts. It was one of the first instances of cannons being hauled into battle by a marching army—hitherto they had been cumbrous and difficult to move and required special gun placements to withstand the recoil of each blast. A new method of casting the cannons whole, like a church bell, however, had allowed for trunnions to be molded on the sides so they could be mounted on a wheeled platform.

The French gunners began to fire quickly on the English formation with great noise and smoke. Stone shots were lobbed from the mouths of the cannons and came bouncing along the turf and smacked into the ranks of the English archers, who were helplessly out of bow range, until they realized they had to either retreat or advance to avoid certain destruction. With a cheer they surged forward in a ragged line toward the infernal guns, killing those gunners who were too slow to flee, and captured the guns before being pushed back by a French cavalry charge.

The resounding boom of the cannons, meanwhile, had attracted the attention of another nearby small French army commanded by the Count of Richemont, who came charging into battle. Kyriel ordered his men to regroup for the assault, but it was futile. With the archers in disarray and in poor formation to use their weapons effectively, the French armies routed the English, killing or capturing nearly all the men while suffering only a few hundred casualties themselves. It was a crushing defeat for the English, leaving them without a significant army in Normandy.

The Battle of Formigny was one of the first conflicts in which cannons influenced the outcome. Over the next few years, French forces armed with mobile cannons pulverized the stone walls of citadels and castles and wreaked havoc on English troops, expelling them from the continent. Old battle strategies and fortifications were rendered useless by the cannons, and the English garrisons surrendered so quickly that when the interminable war officially ended in 1453, the only remaining English possession in France was Calais. Gunpowder weapons in the form of the long-range mobile bombard had superseded the long-range English longbow.

Fresh from his victory over the English in Normandy, Louis XI, king of France, launched a concerted assault against one of his rivals, the formidable Charles the Bold, duke of Burgundy, an overpowerful subject with extensive lands and the most skilled bell casters, and gun makers, of the era. The Burgundians armed to defend themselves against the French king, precipitating an arms race that rapidly perfected gun design. Metal workers strove to improve the mobility of cannons while retaining their firepower. This was accomplished by developing stronger cast cannons that could withstand high pressure needed for heavier and denser iron shot that allowed for the same firepower from much smaller, hence easier to transport, weapons. Cast with trunnions, a six- or eight-foot cannon could be mounted on a sturdy two-wheeled cart and hauled with an army without undue difficulty. When Charles the Bold was killed in 1477, his lands were divided between France and the Hapsburg empire. "As a result," writes military historian John Keegan in *A History of Warfare*, "the French royal house was fully in control of its own territory for the first time since Carolingian days six centuries earlier, and ready to erect a centralized government—supported by a fiscal system in which cannon were the ultimate tax-collectors from refractory vassals—that shortly became the most powerful in Europe."

In the succeeding decades, Louis XI's successor, Charles VIII, further consolidated the French kingdom by annexing Brittany and then marching south to invade Italy and push his claim to the throne of the Kingdom of Naples. His mobile cannons gave him a remarkable set of victories. He marched to the south end of Italy with his swiftly transportable cannons pounding down in days fortresses that had stood for centuries. Keegan writes: "The whole of Italy quaked at his passage. His guns had brought a true revolution in warmaking. The old high-walled castles against which siege-engines and scaling-parties so often failed were hopelessly vulnerable to the new battering instrument." The formidable weapon immensely improved by French and Burgundian metal-workers was poised to quake the foundations of European society.

One seventeenth-century writer, William Clarke, pleased with the power that gunpowder weapons had achieved after several centuries of perfection, wrote: "Let us imagine so many Mars's or Hurcules's of Ancient times, who so undauntedly stood to their Arms; if they now saw the *Artificial Lightning,* and heard the *Thundering Noise,* and experienced the *Power* of our *Guns,* they would presently be amazed, and quake for fear, and be heartless, and would fall down dead at the force of our *Fire-Arms.*" But it wasn't always the enemy who fell down dead from exploding firearms. For centuries the weapons remained notoriously dangerous and fickle, and were apt to explode and shatter without warning. James II of Scotland is perhaps the best-known unintentional victim of the unpredictable power of early artillery. Seeking advantage from the political unrest resulting from the War of the Roses, in 1460 he marched south and assaulted Roxburgh Castle.

While overseeing the firing of a great bombard boldly named "The Lion," he was killed when it exploded while he stood beside it. A stunned onlooker described the scene: "While this prince, more curious nor became the Majestie of any Kinge, did stand near-hand where the Artylliare was discharged, his thigh-bone was dung in two by a piece of a mis-framed gun that brake in the shuting, by the which he was stricken to the ground and died hastily."

Despite these setbacks, and the obvious limitations of the new weapons, mobile gunpowder weapons quickly had a profound impact on the political and military structure of Europe. Fortresses which had stood impregnable for centuries were reduced to rubble within days. The height of stone walls, the greatest asset in a medieval fortress, was no longer an advantage but a liability because its great weight was vulnerable to the percussive impact of cannon shot. The higher the walls and stone towers, the greater the power of gravity to drag them down once the bottom was breached. The historian William H. McNeill writes in *The Age of Gunpowder Empires, 1450–1800*: "The balance of power between central and local authorities was thereby transformed, making whoever controlled the new siege cannon into a sovereign and reducing those who could not afford them to a subjection they had not previously experienced." The old feudal system with its balance of power between regional and central authority was thrown into disarray. The stronghold of a local lord could now be reduced to rubble in hours or days instead of years of laborious and expensive siege. Dissenting opinions from the titular head of state, the king, would no longer be tolerated. A king commanded a far more absolute authority than ever before, compelling servitude from the disparate and far-flung regions of his realm.

Within a few decades of Charles VIII's rapid invasion of Italy, a new style of fortification was being constructed at great expense throughout Europe—dirt-covered walls that dispersed the impact of cannon shot. These fortifications required great numbers of cannons to be effectively defended because they were not very high and were enormously expensive, further enhancing the power of wealthy sovereigns over their regional lords. A narrow window of opportunity existed for all of Europe to be united under one great gunpowder empire, but unlike other regions of the world, the new cannons were never monopolized by any one state, and the opportunity passed before any state could secure hegemony, ensuring a balance of power between dozens of independent quarreling sovereigns.

The incessant internecine small conflicts within Europe rapidly perfected not only the technology of cannons but also the method in which they were utilized. Less than a century after Louis XI used cannons to expel the English from Normandy, gunpowder weapons had changed the way battles were fought: Commanders and tacticians had figured out the most advantageous uses of guns and cannons to augment and support the cavalry and pike-wielding infantry. At the Battle of Marignano in 1515, a French army marching through Switzerland to Italy used the long-range firepower of cannons and arquebusiers to utterly rout a Swiss force of pikeman and knights sent to block their passage through the mountains, killing nearly 22,000 of the 25,000 Swiss soliders, while losing merely 2,000 themselves. Marignano is considered the first example of modern warfare, a turning point in battle strategy that meshed the disparate elements of an army to take full advantage of the tremendous power of guns and cannons. But Marignano not only changed the way battles were fought, it also rearranged the relative importance of the combatants.

For centuries, one of the pillars of the feudal system had been the relationship between military role and social class. Armored knights devoted a lifetime to training with arms and studying the art of combat. Their weapons, armor, and trained warhorses, were extremely expensive, beyond the means of all but the wealthiest. In exchange for their time, the danger to their life, and these great

Cannon technology improved little over the preceding three centuries until the development of smokeless powder in the 1880s.

expenses, knights were accorded a privileged position at the pinnacle of society. With effective gunpowder weapons, however, the greatest champion could be easily shot dead from a distance by a relatively untrained peasant. Knights were no longer as unassailable and feared. As a result, the privileged social preeminence of the noble warrior class began to disintegrate and the feudal social structure, where knightly service was performed by local lords in exchange for their lands and tenure, collapsed. Arming, equipping, and training a knight was time-consuming and expensive, and when he could be brought down by a random gunshot, as Cervantes lamented in *Don Quixote,* it was scarcely worth the effort. "The effect of a cavalry charge," writes Keegan in *A History of Warfare,* "had always depended more on the moral frailty of those receiving it than on the objective power of the horse and rider. And once the horsemen encountered an opponent who could muster the resolve to stand . . . or a weapon that could bring a rider to the ground with certainty, as the musket could, the right of the knightly class to determine how armies should be ordered, and the right to retain an equivalent social pre-eminence, was called into question." Winston Churchill in *A History of the English-Speaking Peoples* wrote of the impact of explosives and gunpowder weapons on feudal European society: "Amid jarring booms and billowing smoke, which frequently caused more alarm to friends than foes, but none the less arrested attention, a system which had ruled and so guided Christendom for five hundred years, and had in its day been the instrument of immense advance in human government and stature, fell into ruins. These were painfully carted away to make way for new building."

The change, not surprisingly, was resisted as long as possible by the mounted aristocratic warrior classes. The Italian knight Gian Paolo Vitelli was so morally offended by the challenge to his station that he ordered all enemy arquebusiers and gunners to have their hands severed and their eyes gouged out. The derision and sneering that had pronounced crossbows and longbows ignoble weapons fit only for cowardly wretches was doubly applied to the new gunmen, who were usually drawn from the lower orders of society. But by the late seventeenth century, gunpowder weapons were ubiquitous and standard in any army. Contemporary commentator William Clarke waggishly remarked on the ascendancy of gunpowder weapons. "Here we may consider not without admiration," he pronounced, "how all other Inventions of War, which had been brought to such perfection by so many succeeding

Ages, should all be disus'd, and vanish at the sight of a *Gun,* except the *Sword,* which is wore as much for fashion as use, a *Pocket Pistoll* being preferr'd to it for security." Social intransigence aside, the perpetual shortage of saltpeter for gunpowder also prolonged the use of handheld weapons in warfare. Pikes and swords, of course, have their applications where a gun would be useless, such as hand-to-hand combat or at sea when boarding another ship, but many soldiers were forced to rely on bows, swords, and pikes into the eighteenth century, when guns would have been preferred.

Change was inevitable, however, if a nation wanted to defend itself or pursue military ambitions. Armored knights slowly became an anachronism, expensive and unruly, yet of little use against a trained force of men with cannons and guns. The Scottish historian Thomas Carlyle wrote that "gunpowder made all men tall." Gunpowder weapons didn't need nobles to shoot them, and they didn't require a lifetime of martial study and practice to master. Height or physical strength counted for little. But these new weapons could not be effectively wielded by undisciplined hordes—their greatest utility came when fired together in volleys. To avoid the danger of arming potentially rebellious peasants with such powerful weapons, firearms required a new type of training that placed discipline, obedience, and teamwork above personal valor and individuality. Peasants were not inherently loyal to the prevailing social hierarchy that relegated them by birth to the lower rungs of society, and commanders sought to produce soldiers loyal to their fellows rather than to a feudal lord or a distant king.

Over time, particularly throughout the seventeenth century, the aristocratic tradition of mounted personal combat was replaced by anonymous armies of peasant gunners, endlessly drilled into mechanical precision and firmly under the command of officers. "Though its organisers might have denied it," Keegan observes in *A History of Warfare,* "we can recognise this as a military slave system, close in character to the Ottoman janissary force, recruited by levy and kept in obedience by harsh discipline and an almost complete denial of civil rights to its members. The style of fighting practiced, that of stereotyped, almost mechanical drill-movements performed in serried ranks, exactly reflected the surrender of individuality its members had undergone." Gone were the days of chaotic mercenary bands decked out in the garish and clashing garments they had seized in plunder, or the colorful knight arrogantly displaying his banner and coat of arms while charging about the battlefield in search of opponents who matched his social standing. Feudalism and all its elaborate and ornate trappings, pretensions of honor, ritualized combat, and rigid social hierarchy were swept away by gunpowder.

The use of the new gunpowder weapons was not restricted to Europe. In 1453, the same year that Louis XI was completing his rapid expulsion of the English from their century-long occupation of northern France, Ottoman invaders broke the defenses of the last Christian stronghold in the Near East, sacked and plundered its wealth, and renamed it Istanbul. They accomplished the feat, which had met with failure on numerous previous occasions over the centuries, with giant cannons. Because of the twin events of 1453, historians consider it as marking the end of the Middle Ages and the beginning of the modern era.

For over a millenium the great city of Constantinople, center of the Byzantine Empire, had been the commercial and cultural capital of the Roman Empire in the East, a wealthy, thriving, and cosmopolitan metropolis of over one hundred thousand that had withstood countless assaults from Magyars, Russians, Saracens, and Turks. It relied on its famous three-walled defenses, a disciplined and large fleet, and the mysterious and deadly properties of Greek fire to preserve its independence. In 1451, a reputedly dissolute and sadistic young man had assumed the title of Sultan

Mehmed II of the Ottoman Empire, a new and powerful force arising in present-day Turkey, and quickly set out in pursuit of his boyhood dream of conquest. He transformed himself into a grim and single-minded tactician, spending the next two years planning for the destruction of the final remnant of the Roman Empire.

In 1453 Mehmed II amassed a mighty force of around two hundred thousand men and a fleet of war galleys to destroy Constantinople. The territories surrounding the famous city-state, in present-day Romania and Bulgaria, had already been subjugated by the Ottomans such that the Byzantine Empire resembled a head without a body. Quickly grasping the change that was resulting from the perfection of cannons in Europe, the new sultan enlisted the services of a Hungarian military engineer and metalworker named Urban, who had apparently been mistreated by the Byzantines, to forge an immense siege train of artillery for his assault. Mehmed's mighty artillery force consisted of a dozen great bombards and nearly sixty smaller mobile cannon. He also ordered the building of a monstrous great bombard named Urban's Bombard (after its builder). The mighty bombard was twenty-six feet long, had a thirty-six-inch bore, and fired a ball of twelve hundred pounds. Nearly two hundred men and dozens of oxen slowly dragged it, and the other artillery, into place on April 6 and prepared to pound down the walls. Urban's Bombard was so large that to pack it with powder and scour the barrel after each shot required hours, and it only fired seven times a day. It also suffered from structural irregularities and after a few days it cracked. But its initial shots and the relentless six-week pounding by Mehmed's other artillery shattered great chunks of Constantinople's walls.

The Byzantine Emperor Constantine XI (Paleologus) knew Mehmed's earth-shattering bombardment doomed him and most of his citizens. His defending army consisted of only eight thousand men; if the impressive walls of his city were breached, they would be overwhelmed by the vast numbers of the invaders. He tried mounting his own, inferior, artillery on the walls to return fire at the sprawling army. But it proved a lame and hopeless gesture because the old fortress walls had been designed for height, and were not sturdy enough to absorb the recoil of cannons. With each blast a portion of their own stone walls cracked and crumbled. There was little the defenders could do to stop the relentless pounding of their city walls, and after six weeks the great gaps were sufficient for Mehmed to order his vast army forward. The city was overrun on May 29, 1453. Constantine, apparently hearing that the invaders had entered the city through a demolished gate, brandished his sword and claimed, "As my city falls, so I will fall with it." He was soon stabbed and killed, and the raping, looting, vandalism, and burning began. Several thousand citizens were killed in the chaos, and tens of thousands of others were sold into slavery. The defeat of Constantinople was the first great artillery triumph against a previously impregnable fortress. It was so successful that in the succeeding years the Ottoman Empire, which would endure until 1924, rapidly embraced cannons and smaller firearms and set out on a military expansion from Syria, Arabia, and Egypt into Hungary, the Ukraine, and the Balkans.

Seventy-three years after the fall of Constantinople, in 1526, Babar, a descendent of the great Mongol warlord Tamerlane, crossed the Himalayas and invaded northern India, hauling with him a vast train of Turkish artillery. Although most of his invading army consisted of mounted warriors, it was Turkish guns that allowed him to defeat a mighty force of a hundred thousand men and perhaps a thousand elephants at the Battle of Panipat on April 21, securing a toehold for the conquest of most of the Indian subcontinent. Although the invaders were mounted warriors and archers in the Ottoman and Persian tradition, Babar was astute enough to recognize the inherent strategic value of cannons. Like Mehmed the Conqueror he hired artisans and metalworkers to cast great siege cannons on location near fortresses he was assaulting. The guns were too huge and cumbersome for swift-riding cavalry to transport. While his cavalry swept the field of foes, his siege guns destroyed their refuge and defense. Babar and his successors established the foundation of the Mughal Empire, which subjugated the Hindu population and ruled most of India, although somewhat loosely in certain regions, for several centuries. The monopoly of siege guns in India, which was never achieved in Europe, brought and maintained a strong central authority at Delhi. And this dependence on cannons established the great saltpeter production in Bihar that was attractive to Europeans in the seventeenth and eighteenth centuries and eventually brought the downfall of the Moghul Empire when it faced the superior guns and tactics of modern European drill armies who sought to secure control over the trade in valuable commodities such as cotton and saltpeter.

Gunpowder weapons shaped the evolution of society in China and Japan as well, but in an entirely different fashion than in Europe or the Middle East. Although gunpowder and the earliest primitive guns originated in China, they never became common or widely used by the military. It became obvious to the ruling elite that a gun in the hands of a common peasant or outlaw could drop an armored noble warrior as easily as a stray dog, and therefore presented an unacceptable challenge to the social order. In both China and Japan, the ruling elite not only had a vested interest in maintaining the existing social order by restricting the spread of dangerous firearms, but they had the means to do it. While Europe and Ottoman Turks shared a frontier, making the development and use of gunpowder weapons indispensable for survival in the clash of civilizations, both China and Japan had the luxury of relative isolation and internal cultural stability which allowed central authorities to restrict the use of gunpowder weapons in their militaries in order to preserve centuries-old martial traditions.

The historian McNeill has noted that "no one in China advocated a policy of equipping infantry with handguns and then training them so well as to allow them to meet and overcome cavalry in the field. A mobile infantry army, recruited from the bottom of society, would have been difficult for civilian officials to control." Because for centuries the principal external threat to China arose from the nomads of the central steppes, mounted warriors who had no access to gunpowder weapons, the Chinese central government could safely pursue a policy of arms control within the military: Crossbows were just as effective as guns at defending fortifications when the attackers were not equipped with long-range artillery. The military was structured such that no single commander could lead a large field army without the consent of the civil authority. Commanders were denied access to firearms and prevented from acquiring the power and influence to demand them. By preserving an equilibrium of power that never allowed any single group to wield too much influence, combined with freedom from external invaders armed with superior weapons, Chinese authorities were able to maintain their prevailing social hierarchy. This style of governance produced no need for expensive and advanced weaponry.

The suppression of powerful gunpowder weapons, however, created internal stability at the expense of a strong military for defense—a policy that was successful for several centuries but eventually exposed the Chinese to the real possibility of foreign invasion when Europeans began landing on the coast. When the coastal regions became the frontiers of contact and conflict with foreigners instead of the steppes, the Chinese military was ill prepared and ill equipped to meet the challenge. During the Opium War with Britain between 1839 and 1841, small numbers of well-armed and well-trained British troops swept away Chinese defenses, starkly revealing how archaic the Chinese military had become during its centuries of isolation and the power of gunpowder weapons against those who fought without them.

In the sixteenth century, Japan had a social structure and hierarchy that roughly paralleled Europe's feudal system, with the samurai being the knightly class who owed allegiance to regional lords and a nominal emperor, and who justified their exalted position near the pinnacle of society by a lifetime devoted to military training in swordplay and a tradition of ritualized personal combat. Honor, duty, and ritual were paramount. Soon after the time of European contact in the early 1500s, gunpowder weapons had a dramatic and revolutionary impact on the history of the nation. It was a time of great conflict between rival warlords and clan coalitions who battled for supremacy. When Portuguese traders showed up on the coast with their primitive arquebuses in 1543, their new weapons were of immediate interest. Within a few years Japanese metalworkers, perhaps the most skilled in the world at making swords, began turning out great numbers of quality handguns and small cannon for the increasingly desperate and violent internecine struggle. In 1575, a force of ten thousand troops, consisting primarily of farmers armed with arquebuses under the command of samurai officers, shifted the balance at the battle of Nagashimo, ensuring the victory of the strongman Shogun Oda Nobunaga over his rival and enabling him to secure political influence over much of the nation. The nature of the victory, where the balance of power lay not with the cultured and noble samurai but with the peasant gunners, disturbed notions of propriety among the ruling class. The traditional role of the samurai was being undermined—usurped by unskilled peasants. Although the samurai were still the commanders, their individual status as great warriors was seriously threatened.

When the quarreling regional warlords of the country were finally brought to heel by the Shogun Tokugawa Ieyasu in 1600, he, with the support of the samurai class, began to restrict the availability of gunpowder weapons. Ieyasu also began disarming the country, eliminating guns and cannons altogether and restricting swords to the traditional military class, the samurai. Throughout the seventeenth century gunpowder weapons, as offensive to the samurai as they were to the European knight, became increasingly scarce, rusting away under government edict. John Keegan puts it succinctly in *The History of Warfare:* "The Tokugawa and their predecessors may have used gunpowder for reasons of *Realpolitik;* once it served their purpose of winning them power, it and all firearms became detestable."

By eliminating potentially destabilizing gunpowder weapons, the sword-wielding ruling class maintained its influence and exalted social position. Since there were no internal conflicts, wars ceased and there was never a need to arm commoners. The central authority was successful in this because they were able to isolate the country, forbidding outsiders to land and forbidding Japanese ships to sail across the oceans. But after two centuries of isolation from foreign contact, the samurai were wholly unprepared for the arrival of American ships in Tokyo Bay in 1854. They soon embarked on a rapid modernization of their military. As with the Chinese around the same time, the rigid social hierarchy and military technology inherited from centuries before with little change was rapidly and painfully deconstructed and rebuilt to meet the new challenges— perhaps even quicker than the collapse of feudalism in Europe.

Though heavily weighted in its most obvious and visual sense to the military, the role of explosives in affecting social change had another more subtle and less direct influence on society. From as early as gunpowder had been perfected, experimenters and curious inventors began searching for methods by which they could harness the chemical explosion for other, more controlled or nuanced, uses. Guns and cannon barrels were the most obvious and practical means of using the pressure from the explosion to launch a projectile. But as early as 1508, Leonardo da Vinci was speculating on yet a further use for gunpowder's explosion: an engine that could be used to lift weights or pump water. In 1673, the Dutch scientist and inventor of the pendulum clock, Christian Huygens, drew up a plan for an engine that could use a controlled detonation of gunpowder to build pressure and drive a piston inside a metal tube. "The force of cannon powder," Huygens wrote, "has served hitherto only for very violent effects, such as mining and blasting of rocks, and although people have long hoped that one could moderate this great speed and impetuosity to apply it to other uses, no one, so far as I know, has succeeded in this; and at any rate no notice of such an invention has appeared." He optimistically postulated that his newfangled device would one day enjoy widespread uses, such as powering mills and propelling carts.

In the following years Huygens's assistant, Denis Papin, built the first model of the gunpowder engine at the Académie Royale des Sciences in Paris, where he hoped that it could be used to mechanically suck water from the Seine River and spray it skyward from Louis XIV's famous fountains. Alas, gunpowder combustion was not ideally suited to the tranquility of the fantastically manicured and tranquil gardens. Not only was it loud, but the explosion was too powerful and most of the gasses had to be released through a system of valves as the piston was pushed out; as the remaining gasses cooled they sucked the piston back, but with an insufficient vacuum because not all the gasses could be evacuated from the cylinder. After years of work perfecting the design, Papin abandoned the scheme as impractical, but not before he was inspired to take the principle in a new direction. Instead of a gunpowder reaction, he wondered about the wonderful expanding property of heated water. "Since it is a property of water that a small quantity of it, turned into vapour by heat, has an elastic force like that of air, but upon cold superventing is again resolved into water, so that no trace of the elastic force remains, I readily conclude that machines could be constructed wherein water, by the help of no very intense heat, and at little cost, could produce that perfect vacuum which could by no means be obtained by the aid of gunpowder."

Papin's primitive steam engine was tinkered with and perfected in the coming years by other inventors such as Savery, Newcomen, and Watt, and the principle was later adopted for gas engines and internal combustion engines. The historian J. D. Bernal commented in *Science in History* on the role of gunpowder on scientific invention. "Ultimately," he wrote, "it was the effects of gunpowder on science rather than on warfare that were to have the greatest influence in bringing about the Machine Age. Gunpowder and the cannon not only blew up the medieval world economically and politically; they were major forces in destroying its system of ideas. . . . The force of the explosion itself, and the expulsion of the ball from the barrel of the cannon was a powerful indication of the possibility of making practical use of natural forces, particularly of fire, and was the inspiration behind the development of the steam-engine." The steam engine, perhaps the single greatest technological invention of its era, and the Industrial Revolution that it enabled, radically transformed the European countryside, the prevailing system of production, and the lives and livelihoods of countless millions of laborers. It destroyed the old social order, insidiously and slowly, in a manner as profound as the toppling of the feudal knight by the first gun.

Alfred Nobel's discovery and marketing of a reliable detonator for nitroglycerin and dynamite in the 1860s heralded yet another new era of explosives that had an even greater impact on society than gunpowder did five centuries earlier, and was appreciated and adopted throughout the world more rapidly. Although dynamite's greatest use was in civil engineering and mining, which consumed the vast bulk of dynamite production, its potential was immediately recognized by most military bodies almost from the instant it was available. To fail to appreciate dynamite's power in the 1860s was as foolish and obtuse as a feudal knight refusing to acknowledge the power of guns and artillery.

Despite the setbacks resulting from nitroglycerin's deadly flaws, which made many jurisdictions leery about the unqualified acceptance of dynamite, the explosive was so useful that within a few years it was being consumed throughout Europe and America, except in one stubborn country. For several years Nobel had been trying to introduce nitroglycerin and then dynamite to French markets but had been blocked by a French government gunpowder monopoly over the manufacture and sale of all explosives, L'Administration des Poudres et Salpêtres, which had existed since before the revolution in the late eighteenth century. When dynamite became available, the influential gunpowder interests persuaded the government to include all new explosives under the monopoly as well, and ban them. It was to have disastrous implications for France and the world.

In July 1870 the proud, aging Napoleon III fell prey to a diplomatic trap and haughtily declared war on Prussia. The dire shortcomings of France's military became quickly apparent, not only in organization and numbers but also in technology. On September 1 or 2, merely a month and a half after the war began, Napoleon III surrendered after he and his entire army were captured at the battle of Sedan. The historian Bergengren has wryly observed in his biography of Nobel: "In its very first phases the French general staff found to its horror that German sappers had hit upon the idea of using the new explosive, dynamite, that France had spurned, for blowing up French forts and bridges." It proved to be a great advantage to the invading armies. A contemporary painting depicts the dejected and slumping Napoleon sitting in a chair near the battlefield in discussion with Chancellor Otto Leopold von Bismarck, who sits straight-backed leaning on his sword in victory, his chin jutting arrogantly upward.

When Napoleon was held captive and his empress fled to England, fearing the wrath of the Parisian mob, a new republic was declared in Paris. The Prussian invaders, however, did not then leave the country as the French had hoped, but continued to encircle Paris and establish a siege of the new government. The French people did not accept defeat as easily as Napoleon III. The young, eccentric, yet highly patriotic son of a southern grocer named Léon Gambetta fled Paris by drifting over the Prussian army in a hot air balloon and proclaimed the Third Republic in Tours. He renewed the offensive and remarkably, according to historian Herta E. Pauli, "took a war that was utterly, irrevocably, shamefully lost, and made the victor win it all over again, the hard way—in battles that not only redeemed French honor but forged a nation out of a disintegrating populace." One of Gambetta's first acts was to secure a loan for the young patriot Paul François Barbe, Nobel's agent and business partner in France, to establish a dynamite factory in the south of the country near Paulilles, which was soon in production.

But it was too little too late. The Prussian military superiority soon crushed French resistance and secured dominance over most of the country and demanded elections to form a government to negotiate peace with them. The terrified people, under the threat of the Prussian military presence, voted in a majority who strongly favored peace at any cost and who sought to restore the monarchy. They officially surrendered on January 28, 1871. In March, however, the besieged citizens of Paris, despite being reduced to eating rats and animals from the zoo, with horrible outbreaks of scurvy during the four-month ordeal, supported a radical group of republicans who proclaimed a government of Paris calling itself the Commune and began a civil struggle against the royalist faction. In April the Prussian army, camped in the hills surrounding Paris, watched bemusedly as the French troops under the new National Assembly began the second siege of Paris and assaulted their own countrymen for weeks until a joint artillery bombardment with the Prussian army crumbled the resistance. The anarchists of the Commune executed hostages, burned public buildings, and used dynamite to explode and destroy public monuments in the city before going down to bloody defeat on May 28.

France's loss in the Franco-Prussian war was so humiliating and so complete, the terms of peace so onerous, that it would have repercussions into the twentieth century. The Cambridge historian Alastair Horne writes in *What If? Eminent Historians Imagine What Might Have Been,* that "Under Bismarck's harsh terms, France lost two of its fairest and richest provinces, Alsace and Lorraine. The nation would never forget. Forty-four years after . . . France would go to war to regain them, bringing the whole world into a new catastrophe. The whole world equilibrium would be fundamentally altered, and a second, even more terrible world war would be fought." Certainly, the use of dynamite by the Prussian army was not the only reason France suffered a crushing defeat, but it undoubtedly contributed to the speed with which they lost and hence the punitive terms for peace demanded by Prussia. The historians Bernard and Fawn M. Brodie write, in *From Crossbow to H-Bomb,* that "the interplay of science, technology, and or men in war becomes increasingly difficult to describe as science becomes more complex. Every simplification is likely to be a distortion." Nevertheless, to the extent that any historical speculation on alternate possible outcomes for world events has validity, France's refusal to adopt the new technology of high explosives, coupled with Prussia's eagerness to embrace them, reverberated over the decades and ultimately played a role in kindling the greatest series of destructive wars ever. If France's defeat had not been so swift and complete, Prussia might not have been in a position to demand both Alsace and Lorraine, thereby altering the series of events that culminated in the First World War.

The civil conflict during the Commune of Paris had shown that dynamite, or something very near it, could be safely and quickly manufactured by relative amateurs in great quantities and used for somewhat antisocial purposes. Bergengren observes: "The outrages committed during the Commune of Paris with, among other things, explosives in the form of the rabble's homemade bombs, had aroused throughout the country a somewhat exaggerated terror of any new explosives." After the war the new National Assembly again banned dynamite and ordered Barbe's and Nobel's factory closed. Dynamite and other high explosives, cheap, relatively easy for individuals to manufacture on their own, and highly portable, have been used as the weapon of social conflict by anarchists, terrorists, and rebels since their invention. Dynamite was not manufactured in France or readily available for sale until 1875, years after it had been embraced by the rest of Europe for both civil and military uses. Dynamite placed great power in the hands of individuals and was feared in nineteenth-century Europe for the same reasons that gunpowder was initially feared centuries before, and it was universally adopted and accepted for the same reasons gunpowder had been—because the fear of not adopting it when neighbors did, which was brought into obvious light by the quick defeat of France during the Franco-Prussian War, outweighed the fear of internal civil dissent. The sale of dynamite was prohibited only in Russia, which was able to keep it under strict government control better than other states, for the same reasons that China and Japan were able to restrict gunpowder centuries earlier, because it was morè isolated.

For most of Europe and America the era of gunpowder ended swiftly and was replaced by the regime of much more powerful dynamite. The pace of industrial change had been rapidly accelerated by the new technology. For five centuries, gunpowder was the only explosive, and the new methods of controlling and harnessing its power in the early nineteenth century were little more than improvements and modifications of the original concept and designs first used in the late fourteenth century. There had been few significant technological innovations, and no significant increase in explosive capacity, for several hundred years. Yet within a few decades of nitroglycerin being marketed, dozens of similar products were available, with adaptations for use as industrial explosives, bombs, and artillery shells and as propellants for guns. In the late nineteenth century, explosives were big business, intellectual property law was in its infancy, and science was riding a wave of unparalleled creativity.

Inventions, Patents, and Lawsuits

The Golden Age of Explosives

Even with such a luxury of patents, the protection in most cases would be illusory. I therefore propose giving the patenting of chemical improvements the name "Taxation of inventors for the encouragement of Parasites."

—Alfred Nobel, ca. 1870s.

Alfred Nobel was the chief progenitor of nearly every advance in the development of explosives during the late nineteenth century, yet he spent more time battling to protect his intellectual property than he did on research. There was just so much money to be made, and Nobel's work and inventions, though truly revolutionary in their near immediate impact on society, were astonishingly simple in concept and very easy to duplicate. Although there were countless other chemists and scientists working toward similar objectives—the demand for new, more powerful explosives was ever escalating and the potential profit was immense—much of their work was loosely based upon Nobel's original discoveries, minor improvements to a fundamentally sound design. Not all competing explosive products were direct copies of his work, but many were. Just to keep abreast of all the possible patent infringements, Nobel had to be ever vigilant in numerous countries on several continents. Within only a few years of the invention of dynamite, there were dozens of nearly identical products, with the only difference being a new stabilizing base for the nitroglycerin, often not as effective or safe as kieselguhr. These explosives were given suitably manly names like Rend Rock, Vigorite, Atlas Powder, Hercules Powder, Cliffite, Rippite, Saxonite, and Lithofracteur (rock breaker).

From as early as Nobel's first trip to the United States in 1865 to market nitroglycerin, he found himself caught up in the wild and heady optimism of the post–Civil War economic boom. He faced stiff public and political opposition from the great gunpowder interests, particularly the Du Pont de Nemours empire. Herta E. Pauli writes: "In the Old World, the story of high explosives would be one of inventions, high finance, power politics, and wars. In the New World it started as one of patents, petty swindles, litigation, and accidents." One of the key rascals was a distinguished southern gentleman named Taliaferro Preston Shaffner, a wandering vagabond with high political connections who had served as an advisor to various European governments on explosives and torpedoes, despite his lack of experience and knowledge. Shaffner instinctively knew that high explosives would be extremely valuable in the postwar United States and had sought to purchase the American patent from Nobel in 1863, just after the Heleneborg disaster destroyed his lab and killed his youngest brother. Shaffner offered ten thousand "Spanish dollars"—a transaction that Nobel, sniffing a rat, turned down. But Shaffner was as persistent as he was shady. He then tried to bribe the U.S. minister to Sweden, James H. Campbell, to obtain through diplomatic pressure the secret for detonating nitroglycerin. When this failed, he returned to America and marketed himself as a torpedo and underwater mine expert for the military and bided his time. He was given an honorary title of colonel and publicly boasted of manufacturing the first nitroglycerin in the United States, which he certainly did not do. At the time Shaffner had no idea how to manufacture nitroglycerin.

Nevertheless, Shaffner's opportunity came in 1865 when he got word that Nobel had applied for an American patent for nitroglycerin and it was soon to be accepted. The original patent application was a disorganized mess, poorly translated from Swedish and reorganized by Nobel's Philadelphia lawyer into the proper format. Observing the loopholes, flaws, and vague language, Shaffner hastily put together several similar patent applications himself and claimed priority and patent infringement by Nobel. The legal proceedings dragged on over the winter of 1865–66 and it was only after months of evaluating the evidence and conducting interviews with witnesses that Shaffner's lawsuit was tossed out. But Shaffner did not despond. He shamelessly went to visit Nobel in New York and offered his services as an advisor.

When this, too, failed, Shaffner continued his tactics of dishonest bluffing and political interference. Several Nobel biographers have suggested that Shaffner was responsible for a piece of proposed legislation that had the potential to frighten Nobel and his associates from ever marketing nitroglycerin in the United States. On May 9, 1866, a bill was introduced by Senator Zachariah Chandler of Michigan to hold manufacturers liable for any direct or indirect deaths that occurred in the making or shipping of nitroglycerin products and made it a crime of "murder in the first degree" punishable by "death by hanging." A week and a half later, on May 17, Nobel sold Shaffner all the patent rights to exploit nitroglycerin for military purposes for the grand sum of one U.S. dollar. The very next day the nitroglycerin bill was amended to lower the standard of responsibility for manufacturers to negligence and reduce the crime to manslaughter and the punishment to "imprisonment for not less than ten years." On June 2, Nobel transferred all his rights to his American patents to a company called the U.S. Blasting Oil Company with Nobel holding a quarter of the shares and receiving a $10,000 payment up front. Shaffner was on the board of directors. Then Shaffner patented his own method for shipping the deadly oil, a simple mechanism that consisted of dual canisters, with the nitroglycerin contained in the inner canister. Under the new Nitroglycerin Act this became the only certified way by which nitroglycerin could be legally transported in the United States. Shaffner then sold his license to the U.S. Blasting Oil Company. In the end, the company never produced much nitroglycerin, merely issuing licenses for others to do so, with Shaffner pocketing much of the cash until it went bankrupt, leaving dozens of angry shareholders who had lost a small fortune. Shaffner then sued Nobel, unsuccessfully, for the return of the original $10,000.

Shaffner was not the only rogue. No sooner had dynamite been invented than there were swindlers and patent infringers springing up all over the country with numerous small enterprises supplying dynamite derivatives, which were never marketed or sold under the name dynamite. California was one of the greatest and most eager markets for explosives because of the construction of railways and the mining boom, and Nobel sold his patent rights for dynamite to the Giant Powder Company, which marketed it under the name Giant No. 1. Soon there were not only direct copies being manufactured small-scale throughout the state, but also, eventually, some surprisingly effective original local improvements. One such improvement originated with the California Powder Works, the chief competitor to the Giant Powder Company. They hired an enterprising young man named Jimmie Howden, who had earlier established a mobile nitroglycerin production outfit at the face of the tunnels of the Central Pacific Railroad in the few years before dynamite was invented, to devise a new product that would not be subject to the dynamite patent. Howden gave them a product he called Hercules Powder. The advertisements for the product depicted a muscled hero boldly astride the bodies of vanquished giants underneath the slogan "We knock them all out!" It was a powerful "active base" explosive similar to Nobel's blasting gelatin and originated at around the same time. Du Pont de Nemours and Company eventually bought out the Giant Powder Company and then ceased declaring that dynamite was a public safety hazard.

Nobel was so sickened by the wild nature of postwar American business—the legal wrangling, unscrupulous dealing, and political manipulation—that he left the country and never returned, eventually selling all his interests in any American companies. "In the long run I found life in America anything but agreeable," he claimed. "The exaggerated chase after money is a pedantry which spoils much of the pleasure of meeting people and destroys a sense of honour in favor of imagined needs." But it was not just in America that his inventions went unprotected and government interference threatened to undermine his business and his sense of fair play.

In Germany, Nobel was unable to secure patents because of the decentralized governmental structure of its numerous states. He had to share the market with dozens of small-scale imitators for years until he drove them bankrupt with his sheer persistence and predatory business tactics, and then organized the others into a noncompeting cartel in 1886, which persisted until the outbreak of war in 1914. In France, as discussed earlier, the state gunpowder monopoly persuaded the government to prohibit the manufacture and sale of dynamite within the country. While in Britain, his problems were compounded by a nitroglycerin explosion in Belgium in 1867 that provoked the British government to implement a ban on the "manufacture, import, sale and transport within Great Britain" of nitroglycerin and any material containing it. A year later when Nobel sought to import dynamite, he was initially denied the legal certification. While this had the noble ring of protecting public safety, it was at least in part the result of the influence of a British government chemist named Frederick Abel, who had long been an advocate for a similar, though weaker, explosive called guncotton—a type of nitrocellulose that was, as we have seen, a more highly nitrated version of collodion, one of the prime ingredients in Nobel's blasting gelatin in 1875. For years Britain had used guncotton more freely than any other country, owing to its early history and to the fact that it had been stabilized and rendered relatively safe by one of the government's own chemists.

Guncotton was discovered by a German chemist named Christian Friedrich Schönbein in the mid-1840s. A short, jovial, plump and amiable man, Schönbein rose from humble origins as the eldest son of a dyer and postal worker to become a professor at the University of Basle, a position he held from 1829 until his death at the age of sixty-nine in 1868. Self-educated in chemistry, physics, mathematics, and numerous languages, the majority of his research at the university focused on oxygen. According to one story, he came upon guncotton as a complete accident while heating a beaker of nitric and sulphuric acids over the stove in his kitchen while his wife was away. The container broke in the heat and acid sprayed over the floor. Schönbein, rushing to cleanup the mess, grabbed a kitchen towel, sopped it up, and placed the towel near the hearth to dry, where it began smoking and then exploded. Voila! Guncotton was discovered. It had the texture and consistency of rough, slightly brittle cotton wool and was more powerful than gunpowder but less powerful than dynamite. A more likely scenario is that he came upon it while researching oxygen in his laboratory.

Regardless of how he first came upon the explosive properties of guncotton, Schönbein immediately grasped the potential commercial and military use for the volatile substance. After conducting a battery of experiments he declared that the "curious" new explosive was "in every respect superior to the best powder" and he wrote to some colleagues in England to see about patenting and marketing the substance. "The manufacture is not attended with the least danger," he claimed, "and does not require costly installations. In view of these properties we cannot doubt that this explosive cotton should rapidly find a place in the pyrotechnic arts and especially in war vessels." In August 1846, the same year Sobrero discovered nitroglycerin in Italy, the naïve and optimistic Schönbein boarded a steamer and crossed to England where he gave a successful demonstration at Woolwich with Queen Victoria in attendance and then trundled down to Cornwall at the behest of the mighty mining firm John Taylor and Sons. In Cornwall, he impressed the conservative and skeptical miners, and John Taylor eagerly purchased the English patent rights that same year.

Production of guncotton began immediately, with the product being advertised as six times as powerful as gunpowder. Just how powerful guncotton really was, was soon to be revealed. Taylor struck a deal with the gunpowder firm John Hall and Sons to transform their facility near Faversham, Kent, into the first guncotton factory, with exclusive production rights for three years. Schönbein was to receive royalties. On July 14, 1847, however, the entire facility was leveled by a terrific explosion that tore the roofs off buildings a quarter mile away, and, like an earthquake, shook the foundations and shattered the windows of houses in the town of Faversham, a mile distant. "Eighteen persons were killed by that explosion, ten only could be recognised, the remainder were literally blown to atoms and scattered with the materials in every possible direction. One other person who inhaled the fumes of the acid, and who acted incautiously in not attending medical advice, also died on the evening of the explosion. Of the survivors, fourteen in number, who suffered dreadfully by broken limbs, contusions, and being burnt by the acids, one has since died and we fear one or two more will hardly recover." The explosion ultimately killed twenty-one workers and injured dozens of others. The company wrote to Schönbein to relate the devastating news, and to shift some of the responsibility onto Schönbein's shoulders. He had, after all, promised them that the process was safe and cheap.

Schönbein was horrified, but he speculated that the problem arose from human error in the drying process, and continued to believe that guncotton might still be manufactured safely and profitably. He pursued the commercial exploitation of his discovery in Europe, entering into agreements with other manufacturers in France and Austria with promising results, until these budding enterprises were also destroyed by explosions at Vincennes, Le Bouchet, and Vienna, thereby ending the optimistic rise of guncotton as an alternative to gunpowder. For over a decade and a half, guncotton was ignored. No company sought to manufacture it, because it was just too dangerous. Twenty years later, however, in the mid-1860s, the Great Eastern Chemical Works in Sussex revived it after slightly improving the safety of the manufacturing process by introducing an additional process to cleanse the impurities in the cotton. It enjoyed some small-scale success in the Cornish mines, where the ropelike coils were stored underwater before being stuffed into boreholes, covered in gunpowder, and detonated. Guncotton's scope remained very limited, however, until 1865 when Frederick Abel patented a new method for its production, producing a safer, less volatile form of guncotton.

Abel, like his contemporary Nobel, was a shy, somewhat awkward bachelor. He was thirty-seven years old in 1864 when he left his posting as professor of chemistry at the Royal Military Academy at Woolwich and became the chemist to the Office of Ordinance, responsible for the development and advancement of gunnery and explosives technology for the British military. He remained at the War Office for most of his distinguished career and was eventually knighted for his services. It was an odd career path for the son of a professional musician, whose grandfather had been a court painter. Abel was himself a talented performer and musician but turned his mind toward chemistry when he was accepted as a youth to the prestigious and newly founded Royal College of Chemistry. In 1865, he hit upon the idea of combining two adaptations being proposed for improving the safety of guncotton: pulping the cotton like paper and then washing it extensively to remove impurities. It was the impurities, as in nitroglycerin, that degenerated over time and caused guncotton to explode unexpectedly. After Abel's new technique was perfected, guncotton was exceptionally stable, such that it could be pounded, drilled, and sawed with woodworking tools and shaped for use in artillery bombs, land mines, and torpedoes. Abel also devised a cheaper, less powerful version that was a blend of guncotton and barium nitrate, which he called Tonite. For a brief while Tonite competed with dynamite in England and enjoyed small-scale use in America. Due to the prominence of Abel's position within the government, guncotton was considered so safe that British trains would transport it with only the same precautions used for gunpowder, while dynamite was banned. Within England guncotton enjoyed a respectable following, although it never took off globally because of competition from dynamite.

Although dynamite was a superior product, years passed before it was manufactured or used in Britain. Abel used his influence to keep dynamite out of the country, thereby preserving guncotton as the prime industrial explosive. When Parliament sought expert advice to help draft the Nitroglycerin Act, Abel was selected as the official consultant, despite the fact that he had a financial and personal interest in keeping dynamite out of the country. He easily convinced Parliament that guncotton, though less powerful, was far less dangerous than dynamite, which resulted in the 1869 prohibition on "the manufacture, import, sale, and transport of nitroglycerin and any substances containing it." Guncotton was in fact a very similar substance to dynamite, and suffered from the same problems of volatility, so the regulation served mostly to prevent

Shown are the remains of the Faversham guncotton plant after the terrible explosion in July 1847 (from the London Illustrated News*).*

dynamite from being a competitor to the domestic product. Nobel pointed out in a letter to the British government in 1870 that 560 tons of dynamite had so far been produced in Europe and America without incident—a remarkable safety record for an explosive equivalent to 2,800 tons of gunpowder. "The distinguished advocate for guncotton in England and chief adviser of the House of Commons on Dynamite matters," Nobel wrote, "therefore appears to have exaggerated the danger of the substance. Or, if it is so highly dangerous as he reports it to be, it is certainly a wonderful 'run of luck' to have no accident on such large quantities. . . . A simple reference to statistics will show that the use of firearms for play is productive of incomparably more accidents than this substance which is a great and valuable agent for the development of our mineral wealth." Nevertheless, it took several years to overcome these legal obstacles. It was only in 1871, after pressure from industry, that a special permit was granted for the manufacture of dynamite in England. Owing to Abel's public pronouncements of dynamite's danger, when the ban was lifted Nobel could find no financial backers in London. And so it was that Scottish financiers supported the first dynamite factory in the British Empire.

The British Dynamite Company, Ltd. was established in 1871, with its head office in Glasgow and its manufacturing facility on the distant windswept western coast of Scotland at Ardeer. "Picture to yourself everlasting bleak dunes with no buildings," Nobel wrote, describing the site. "Only the rabbits find some nourishment here ... the wind always blows and often howls, filling the ears with sand which also drifts about the room like a fine drizzle. . . . A few yards away the ocean begins, and between us and America there is nothing but water, a sea whose mighty waves are always raging and foaming." At the celebratory inaugural meeting of the board of directors in Glasgow, Nobel delivered a speech both optimistic and revealing of an underlying arrogance: "Well, gentleman," he announced, "I have given you a company that is bound to succeed even if there is the grossest mismanagement on the part of the directors." No doubt there was a smattering of polite applause.

Because of bad press, not the least emanating from Abel's highly publicized criticisms, railways refused to ship dynamite, and the company had to establish its own fleet of transport steamers and cargo boats that transported dynamite to its markets, which were also used to import raw materials, kieselguhr, and glycerin from Germany and saltpeter from Chile. Even after the Nitroglycerin Act was repealed and replaced by the Explosives Act of 1875, dynamite was singled out by certain port authorities as having to undergo peculiar safety precautions, such as being unloaded onto a huge carpet under an awning, or encasing horses' hooves in flannel shoes, to prevent sparks, presumably. Despite the seemingly bizarre requirements and Nobel's uncharacteristic hubris toward his directors, the British Dynamite Company (later called Nobel's Explosives Company, Ltd.) did indeed succeed, becoming the largest explosives factory in the world and delivering dynamite and later blasting gelatin throughout the British Empire.

While the use of dynamite was delayed in Britain by government interference, it was eagerly embraced in Australia. The rapidly growing mining industry in the vast hinterland had an ever expanding and insatiable appetite for explosives. By the 1880s, despite the nation's low population of about two million, exports to Australia grew from 14 percent of the UK market to more than 50 percent in 1878. By 1883 the country was consuming over 50 percent of the dynamite used in the British Empire. Most of Australia's dynamite and blasting gelatin was imported from the giant explosives factory in Ardeer. The first local production of a dynamite-like substance in Australia was organized by a patent-infringing German scientist named Friedrich Krebs, who first sold his wares in Britain in the early 1870s under the trade name Lithofracteur until he was forced to shut down his operation after years of litigation with the British Dynamite Company. The crafty Krebs reluctantly packed up his operations in England and cast his eyes about for a suitable fertile ground to reestablish his company. Australia seemed ideal: Not only was it undergoing a mining boom but it was sufficiently distant that no other producers had yet set up shop there and it was about as far as one could get from the litigious overcrowded markets of Europe. Krebs established a company called Australia Lithofracteur Company and built a factory along Kororiot Creek, near Deer Park outside Melbourne, which operated on the sly, competing directly with the British Dynamite Company for several decades before being bought out in 1898.

———

"It has always been more difficult to make a good propellant than a good high explosive," writes the chemist and historian G.

I. Brown in *The Big Bang: A History of Explosives*, "just as it is more difficult to sing softly or to dance slowly." So it was that after gunpowder had been swept aside as the primary explosive for military and industrial use, it was still the exclusive propellant used in guns and artillery well into the 1880s, thirty years after nitroglycerin and guncotton had been adapted for commercial use. Gunpowder's inherent properties were a cause of frustration, especially for the military. It was difficult to transport and store, it was susceptible to fouling when damp, and it produced noxious smoke. If a more powerful alternative could be developed, the caliber of guns could be reduced, making them lighter and more easily transportable. And the weight of the ammunition could also be reduced, allowing a soldier to carry more on his person. A smokeless, powerful alternative to gunpowder that was easy to store and transport would be worth a fortune.

Nitroglycerin, guncotton, and blasting gelatin all burned too rapidly and too unevenly for use in guns and cannons. "Ideally," writes Brown, "a propellant must undergo a rapid, but regular and controlled, burning so as to maintain a steady pressure on the projectile while it is in the barrel of the gun. It must not be able to detonate within the barrel or it will shatter or damage the gun; it should be free of smoke and flash and should leave no residue; it should be completely burnt out as the projectile leaves the muzzle; it should be easy to set off, easy to store and to transport; moisture and temperature change should not affect it unduly; and the gases it produces should not be corrosive. All this is a very tall order." Scientists and inventors had been searching for a powerful smokeless propellant throughout most of the second half of the nineteenth century without success. But by the late 1880s, several individuals were rapidly approaching a breakthrough.

The first person to come to market with a smokeless powder was a French chemistry professor named Paul M. Vielle. In 1886, he announced the creation of Poudre B (for *blanche*, because it was white; regular gunpowder, which was black, was known as Poudre N, for *noire*). Vielle's new powder was a combination of two forms of nitrocellulose: guncotton (insoluble) and collodion (soluble) with ether and alcohol. It was nearly smokeless and about two times as powerful as gunpowder. The French military immediately adopted it, partly because of Vielle's political contacts. A few months later, Nobel announced his own version of a smokeless powder. He called it ballistite, or Nobel's blasting powder.

Like much of Nobel's work, little is known of the specific method by which he came upon his discovery. His records show that he had been working intermittently to create a powerful smokeless powder since at least 1879, and surely he experimented with hundreds of possible combinations of ingredients in varying amounts before settling on what he considered the perfect blend for power, stability, and ease of manufacture. We don't know Nobel's thought process because he kept no narrative of his trials, and we don't know all the substances he experimented with or the temperatures and catalysts he used in his laboratory because he kept no detailed records. Certainly it involved innumerable hours over countless days shut up in the stale atmosphere of his laboratory heating beakers of dangerous and volatile acids over bunsen burners and then mixing them together and adding other substances to see what happened, just to develop a prototype. The explosive capacity of each new blend of substances then had to be tested at a blasting range outside the city. Then followed months of work with his assistant chemists perfecting the safest, most predictable, and least time-consuming method to produce the new substance.

The secrecy surrounding much of Nobel's work, and the work of other scientists in the field of explosives and powders for military applications, then as now, derived from the astounding value new discoveries were worth on the open market and the need to protect secrets from competitors. Unlike in other branches of science the emphasis in explosives research was on preventing others from duplicating, or improving upon, one's discoveries rather than facilitating understanding and the advancement of knowledge. They were working on highly valuable substances that were considered industrial secrets, not abstract or theoretical scientific problems. Success came from producing a patentable product.

Nobel's ballistite was a combination of equal parts nitroglycerin and nitrocellulose "of the well known soluble kind," with camphor. "Celluloid, as a rule," Nobel wrote in his 1887 patent application, "contains nitrated cotton to approximately two-thirds of its weight, but owing to the camphor content and the substance's compact consistency, celluloid's combustion, even if fine-grained, is far too slow to make it suitable as a propellant for projectiles. By substituting nitroglycerin, wholly or in part, for camphor, it is possible to produce a kind of celluloid with sufficient consistency to be formed into grains and which, on being loaded into firearms, burns with the subdued rate of combustion." He also noted that: "it generates greater power, it leaves no deposit, and it is smokeless or very nearly so." Like Vielle's Poudre B, ballistite was pliable and could be shaped and sized like dough so as to control the rate of burning, and the grains of the powder could be shaped to fit both short-barreled handguns and mighty cannons. It was a powder that burned so rapidly, and with such completeness, that it gave off no perceivable smoke. In addition to being smokeless, it was exceptionally powerful, exerting over twice the strength of gunpowder in guns. Ballistite was the most scientifically advanced and eminently useful explosive yet created.

Ballistite was a truly revolutionary chemical discovery. In the world of high explosives, it represented the zenith of a half-century trajectory of astounding creativity that swept the last remnants of gunpowder's supremacy aside. A smokeless powder that could be effectively used in guns and cannons without corroding the interior of the barrel was a major technical advance that would radically alter military tactics—artillery and guns could be fired without revealing the location of the men doing the shooting. Explosives could be detonated without a great cloud of noxious smoke. The stakes were huge for the rights to exploit this revolutionary discovery. But despite ballistite being Nobel's greatest technical achievement and source of personal pride, it was to bring him only disappointment and sorrow.

When Nobel triumphantly offered his ballistite to the Administration des Poudres et Salpêtres in France, he was rebuffed. Though ballistite was a superior product, Vielle's political connections and influence had ensured that his Poudre B was rapidly endorsed as the exclusive powder for the military. Furious that he had been beaten by mere months, and believing that Vielle's powder was only adopted because of politics, Nobel scathingly, yet unjustifiably, wrote that "for all governments a weak powder with strong influence is obviously better than a strong powder without this essential complement." When the French turned down ballistite, Nobel turned to other markets, and by 1889 one of his plants in Italy was producing three hundred tons for the Italian military. The following year he licensed the Italian government the patent rights. Despite having turned down ballistite, French authorities were deeply offended at what they considered an act of treason by a resident foreigner (Italy was considered a potentially hostile nation). Despite his having lived in France for seventeen years, the French press launched a frenzied, cruel, and largely unfounded assault on Nobel's character, accusing him of treasonous espionage—of stealing his recipe for ballistite from the laboratories of the Administration des Poudres et Salpêtres. He was vilified and threatened with imprisonment. His laboratory was roughly searched and shut down. Supplies of his various experimental powders were seized. And his gun-testing range was locked and the permits revoked for further experiments. He was prohibited from ever manufacturing ballistite in France. At the same time, Nobel was caught up in a financial mismanagement scandal involving his French business partner Paul Barbe. Harassment by the French government was not Nobel's only stressful concern during this difficult period. His older brother Ludvig died in 1888 and he was also separating from his mistress, Sofie Hess, after nearly fifteen years.

Most galling to the aging Swede was that he was ultimately forced to flee Paris for Italy under a cloud of public displeasure and claims of espionage and intellectual theft. In 1891, Nobel reluctantly packed up the necessary possessions from his mansion, along with any laboratory equipment and supplies that had not been seized, and moved to San Remo on the Riviera di Ponente in Italy, overlooking the Mediterranean. The move distressed him greatly. "It is pure chicanery," he wrote to a nephew, "but since they threaten to put me in a fortress, which would have the disadvantage of further spoiling my digestion, I cannot deny or resist the prohibition. . . . I was in the middle of some very interesting problems which will have to be put aside; it is not at all easy to move my laboratory abroad, quite apart from the expense." It was a very distracting period of time in Nobel's life—distractions that were to cost him a small fortune and usher in one of his greatest personal disappointments. His feeling of betrayal, of being morally wronged and denied credit for a scientific discovery that he believed was his, never left him. It perturbed him greatly for the rest of his life.

Italy was not the only country interested in smokeless powder, and while the French were not immediately revealing the secret of Poudre B, Nobel was peddling ballistite to any buyers. In 1888, the British government appointed a special explosives commission to ascertain the nature of Vielle's and Nobel's discoveries. If a smokeless powder had indeed been invented, Britain would need to have access to it to retain military competitiveness. The mandate of the commission was broad: "to investigate new discoveries, especially such as affected the use of military explosives, and to submit to the War Office proposals for the introduction of any technical improvements in the field." Frederick Abel, Nobel's nemesis from two decades earlier, was on the commission. The two former adversaries over dynamite had been on fairly good terms regarding technical matters, exchanging letters and occasionally meeting in Paris or London. Also on the commission was James Dewar, an influential Scottish physicist and a close colleague of Abel's, who had also been in contact with Nobel over technical matters. When they requested detailed information on production and samples of ballistite, Nobel readily complied. Sometime in 1890, Nobel's Explosives Company in Scotland obtained the patent for ballistite and offered the revolutionary product to the War Office, only to be informed that they had already acquired a patent for a smokeless powder patented by Abel and Dewar, called "the committee's modification of ballistite," or cordite. While looking over Nobel's early notes on ballistite, the two astute chemists noted a problem with the original formula and made a few minor changes before secretly taking out a patent for the new substance and quickly having it adopted by the British military before informing Nobel or his associates.

Cordite differed very slightly from ballistite. It had a higher proportion of nitroglycerin, which made it easier to ignite; it used vaseline instead of camphor, which was a better lubricant and stabilizer; and it contained guncotton (insoluble nitrocellulose) in place of collodion (soluble nitrocellulose). Erik Bergengren writes: "The attempts to settle the matter amicably with the purposeful and self-assured opponents, who stood in such favour in high places, proved futile." Nobel and Abel were both proud and stubborn, and they both believed they were in the right. Nobel's Explosives Company immediately launched a lawsuit over patent infringement that dragged on for years through the Chancery Division Court in 1892, to the Appeal Court, and finally to the House of Lords in 1895, which finally put an end to the exhaustive and expensive quarreling when it ruled against Nobel on a technicality in his original patent application. Nobel's Explosives Company was ordered to pay court costs. It was a bitter defeat that was only mitigated slightly by the perceptive remarks of Lord Justice Kay. "It is quite obvious that a dwarf who has been allowed to climb up on the back of a giant can see farther than the giant himself. . . . In this case I cannot but sympathize with the holder of the original patent. Mr. Nobel made a great invention, which in theory was something extraordinary, a really great innovation— and then two clever chemists got hold of his specifications for the patent, read them carefully, and after that, with the aid of their own thorough knowledge of chemistry, discovered that they could use practically the same substances with a difference as to one of them, and produce the same results one by one." Abel and Dewar may have strayed from a strictly moral path, but they were technically and legally in the right. While the impetus came from Nobel's confidentially submitted information concerning ballistite, cordite was sufficiently different to warrant a distinct patent. "Lady Justice," Nobel wrote to his business partners around the world soon after the failure of his lawsuit, "has always had paralysis of the legs and has therefore been unconscionably slow, but now that she has had a knock on the head as well, she appears too mad even for a mad-house . . . I can afford to be indifferent to the pecuniary side of the case, but I cannot get over my intense disgust with the shabbiness displayed."

Nobel had always been sensitive and naïve enough to feel disgust at swindlers and underhanded dealing. Although he became fabulously rich—the great late-nineteenth-century boom in explosives stemmed directly from his discoveries and he amassed a personal fortune so large it could have lasted several generations of indulgent living—each and every time he felt he was cheated of something, he took it as a personal attack, and brooded and stewed over it until it began to destroy his health.

Nobel was always in the fore, on the cusp of leading discoveries, and he certainly had the corporate infrastructure to quickly and

Frederick Abel, the British chemist who helped exclude dynamite from the British Empire for over a decade and who was embroiled in a lengthy lawsuit with Nobel over priority in the invention of smokeless powders in the 1880s

efficiently bring his new products and improvements to market. But he was not always first, and others had legitimate claims to their discoveries, however similar. In those heady days when high explosives were one of the fashionable fields of research, with numerous exciting new discoveries and advancements and great potential profits, no single individual, corporation, or country was going to get a monopoly. There were too many competing interests, with national security, personal pride, and greed all playing a role in propelling the research and discovery of new explosives. Once an idea was in the air, so to speak, as with the detonator principle, it was relatively easy for chemists to experiment and tinker with known explosive substances to fine-tune them for more subtle or new uses. Nobel wasn't really so hard done by, and one can't help but see a bit of the spitefulness of a sore loser in his pious claims of being beaten by mere politics in France and England. In the end, cordite predominated throughout the British

Empire, Japan, and several countries in South America; Poudre B found favor in France, Russia, and the United States; while ballistite was used chiefly in Italy, Germany, Austria-Hungary, Sweden, and Norway. Nobel's Explosives Company eventually produced both cordite and ballistite (although owing to the ill will generated by the lawsuit, the company received no significant British government contracts for over a decade afterward). The company agreed to pay Nobel a half portion of the royalties from every batch of cordite, so in the end Nobel did profit slightly from it.

There were a great many variations on the key explosives developed by Nobel and others in the second half of the nineteenth century, and it is nearly impossible to present an accurate list of all the different products and brands, some patent infringing, others sufficiently novel, that were developed or were in use by the end of the century, but it surely numbered in the hundreds. It took nearly twenty years to discover a method to predictably detonate nitroglycerin. Then, in rapid succession, came dynamite and all its variants, imitators, and improved versions. Ten additional years brought the even more powerful blasting gelatin and its imitators and variants, such as Atlas Powder and Hercules Powder, that could be detonated underwater. Another decade brought powerful smokeless explosives and powders that could also be used as propellants in guns, thereby completing the eclipse of gunpowder. The quantity of explosives being used in the late nineteenth century for civil engineering projects and by the military was enormous. Although there are no precise figures detailing this astronomical increase in the consumption of explosives, and the corresponding increase in the demand for the age-old ingredient that was as crucial for the new high explosives as it had been for gunpowder, it was sufficient to far exceed the capacity of the traditional sources of this raw material. Salpeter and nitrates were becoming scarce again.

The Guano Trade

The Toil for Chilean Saltpeter and the War of the Pacific

Peru is full of people well appareled and of civil behaviour. It hath many mines of gold and more of silver, has also a great store of copper and tin mines with abundance of saltpeter and brimstone to make gunpowder.

—Lopez Vaz,
Portuguese traveler, 1586

Running northward along the Pacific coast of South America for thousands of miles, from the Antarctic to the equator, is a steady stream of cold water. Cutting a swath through the warm, sunlit waters south of the equator, the chilly upwelling of turbulent water carries with it great quantities of decomposed vegetable and animal matter, photosynthesis nutrients for multitudes of minuscule free-floating plants called phytoplankton. The phytoplankton in turn are nourishment for vast numbers of small free-floating animals called zooplankton (and other tiny filter feeders), which are in turn ideal food for fish. Vast numbers of fish such as anchovies and herring thrive in the Peruvian Current. The fish are food for millions of seabirds such as the white-breasted cormorant, the gray pelican, and the white-headed gannet.

Over the centuries, incalculable generations of these seabirds have feasted on the seething schools of fish and nested to raise their young on remote islands free of natural predators and along parts of the mainland coast. The sheer numbers of seabirds created a great deal of dung. With each bird producing perhaps twenty grams of droppings each day, the millions of birds produced an incredible eleven thousand tons of excrement per year, which eventually filled every cave and crack on the islands, compelling the birds to burrow their nests into the accumulation of their own feces. The local peoples called the excrement "huano" (spelled "guano" in Spanish and then mispronounced to include the "g" by English speakers). The upper layers were ill-colored and yellowish and produced a horrid stench, while the lower layers became hardened and crusty under the weight of each successive year's bird by-product.

One of the first Europeans to observe the unusually large quantities of accumulated guano on the coast and islands was the German explorer and naturalist Alexander von Humboldt, who wandered South America for nearly five years in the early nineteenth century. Although Humboldt could not foresee guano's future military and agricultural value, which fueled Peru's prosperity and export market for over four decades, it did lead his meandering and curious mind to ponder just how it had accumulated in such vast quantities in the first place. A brief visit to an island off the coast of Peru, which was so covered in guano Humboldt had difficulty estimating its depth, suggested to him that it hadn't rained there for centuries, despite the clouds he saw on the horizon. After measuring the surrounding ocean temperatures, he speculated that the unusual phenomenon of the exceedingly dry climate was caused by the particularly cool ocean current that chilled the air above it. The stream of cold water is now known as the Humboldt or Peruvian current. When the coastal air blew over the islands it warmed, increasing its capacity to absorb moisture, which it sucked from the atmosphere. Instead of raining, the clouds floated east and dumped their load at the base of the Andes Mountains, leaving a strip of coastline and nearby islands virtually bereft of precipitation, one of the driest regions on earth. It was so dry on these islands and in a region of the coast known as the Atacama Desert, a zone roughly along the Tropic of Capricorn, that no rain might fall in a person's entire lifetime. The houses of the few natives who lived there were constructed without roofs.

In the absence of precipitation the guano was continuously deposited, year after year for centuries, never evaporating or washing away. By the nineteenth century it had smothered the barren, windswept, and uninhabited islands under a thick crust of between 100 to 150 feet of guano. "Year after year," writes R. E. Coker in *National Geographic Magazine*, "the guano is laid down beneath a clear, dry atmosphere, the deposit bakes in the sun and its most valuable components are imprisoned for an indefinite period." Because it never evaporated or washed away, the Peruvian guano from this region had a particularly high nitrogen and phosphate content. The local Moche people had been using it as a fertilizer from as early as 500 A.D., and later the Quechua continued the practice during the Inca Empire from around 1200 A.D., particularly for use on extensive potato plantations. After it was scraped from the islands and shipped ashore, the guano was sprinkled on irrigated fields to improve crop yields in the bleak and harsh environment of coastal Peru. Guano was so valuable under the Incas that to avoid internal conflicts the deposits on the Chincha Islands were officially apportioned between the various provinces. Seabird nests were protected by law, and disturbing them was made punishable by death. An old Peruvian proverb claimed that "guano, though no saint, works many miracles."

Guano had another important use. Because of its high nitrogen content, the guano was used to make saltpeter for gunpowder beginning in the mid-sixteenth century under Spanish rule. It was never ideal for saltpeter production, however, because it required too much refining. A similar nitrogen-rich product called caliche from the nearby South American mainland, more easily converted into saltpeter for gunpowder and nitric acid for explosives, would later pose a challenge to the guano trade. In the 1820s, however, a newly independent Peruvian government was testing its value as an export commodity. The first shipments of guano were exported to the United States and Britain in the 1830s and the quantities shipped increased rapidly when guano's ability to rejuvenate depleted soils and restore crop yields became widely known. By the 1850s guano was in great demand as a fertilizer for farms in the eastern United States, Britain, and France, and the price had skyrocketed. It was far more potent than manure and seaweed or other commonly available fertilizers, and guano soon enjoyed a worldwide reputation. Demand always outstripped supply during the mid-nineteenth century.

The guano trade was a highly regulated industry under the monopoly control of the Peruvian government, which licensed the export of the commodity to a handful of politically well-connected firms which, despite the high fees paid to the government for the right to ship the guano, enjoyed extraordinary opportunities for profit in a global market that seemingly had an insatiable capacity to absorb its product. The great mounds of guano were transported from South America around the treacherous and turbulent waters of Cape Horn in licensed and registered fleets of speedy clipper ships (the once proud and majestic vessels were compelled to enter the odious guano trade during a glut of shipping capacity in the 1850s and with competition from slower-moving transport ships). The vast bulk of the guano was shipped to England, France, Germany, and the United States, but by the 1850s Spain, China, Australia, and even India were markets for the remarkable fertilizer that proved cost effective (even at its high price relative to other fertilizers) to those who could afford the initial up-front outlay. In the 1850s nearly half a million tons of guano were shipped to the United States, while over one and a half million tons ended up in Europe. By 1860 imported guano made up nearly 45 percent of the commercial fertilizer sold in the United States.

The cost of guano continued to rise along with demand, to the point where it was beyond the ability of smaller or poorly financed farmers to afford it. By the 1850s the price of guano was attracting political attention from the American government, which sought in vain to negotiate a treaty with Peru to circumvent the British firm, Anthony Gibbs and Sons, which held the monopoly rights to ship guano to the United States. Although the Peruvian government actually received nearly 65 percent of the gross proceeds of the guano trade, sentiment in America was that the high price was caused by Gibbs's monopoly. Anger and resentment, and the dream of easy money, spilled over in 1852 when a New York entrepreneur named Alfred Benson nearly succeeded in persuading the U.S. Navy to "defend" his ships while he loaded guano from the Lobos Islands, a handful of small guano-laden islands claimed by Peru that Benson asserted were in international waters. Fortunately, an international outcry and calmer heads prevailed. But the increasing demand led to an international search for more guano deposits.

By 1856 the United States government passed the Guano Island Annexation Act, which gave adventurous American entrepreneurs the "exclusive right of occupying said islands, rocks, or keys, for the purpose of obtaining said guano, and of selling or delivering same to citizens of the United States" at a set price substantially lower than the price charged by the Peruvian guano monopoly. "Whenever any citizen of the United States discovers a deposit of guano on any island, rock, key, not within the lawful jurisdiction of any other government," the act claimed, "and not occupied by the citizens of any other government, and takes peaceable possession thereof, and occupies the same, such island, rock, or key may, at the discretion of the President, be considered as appertaining to the United States." Over the next eight years, until Lincoln suspended the act during the Civil War in 1863, eager American guano hunters staked claims to over sixty tiny islands in the Pacific and a half dozen in the Caribbean. It is for this reason that the United States still owns numerous tiny islands and atolls all over the Pacific which in the nineteenth century were collectively known as American Polynesia. Other nations, particularly France, were also scouring the Pacific for guano islands.

It is a peculiar fact of nature that the same region that was blessed, or cursed, with the greatest supply of quality guano was also endowed with a similar substance that proved to be guano's chief competitor—deposits of caliche, a guano-like substance that lay in the barren, sparsely populated Atacama Desert, the ill-defined no-man's-land that formed the border between Peru, Bolivia, and Chile. As early as the seventeenth century, a cottage industry run by native workers on the eastern slopes of the Andes Mountains above Lima was purifying raw caliche, of which 50 percent was salt and mud, and then converting it to potassium nitrate for the production of gunpowder for use in the local silver mines. The caliche differed from common saltpeter because it was mostly sodium nitrate instead of potassium nitrate (although some potassium nitrate was found blended with the caliche). The sodium nitrate could not immediately be used in the manufacture of gunpowder because it absorbed moisture, making the powder damp and hard to ignite. It was difficult to convert the caliche to saltpeter on a large scale in the desert without wood ashes, which were obviously in short supply in a region far from forests. Small-scale local producers relied on a lye solution made from burnt cactus to transform the sodium nitrate to potassium nitrate in a process similar to the craft production of saltpeter in Europe and India. The historian M. B. Donald, writing in the *Annals of Science*, explains: "The Indians removed the crust of the earth, crushed it into small pieces, and allowed it to soak for 24 hours, in water contained in a cowskin shaped like an inverted cone, with a wooden plug at the bottom outlet. The liquid was removed and boiled in kettles for one hour then left standing in vessels for 24 hours, when the nitrate separated as dirty coloured crystals. These were purified by redissolving in water and adding beaten up egg, which was followed by straining through cloth into glazed earthenware receptacles." Later the process became more formalized, but it was still undertaken on a small scale for most of the seventeenth and eighteenth centuries. Initially, the caliche was more expensive to convert into saltpeter than guano because of the necessity of transporting it from the desert to the coast.

By the late eighteenth century, the use of caliche to produce saltpeter for gunpowder was a growing industry, with some of the product even being shipped to Spain during the Napoleonic Wars to alleviate shortages. And by the 1830s a small but steadily increasing quantity of Chilean caliche nitrates were being shipped to Europe as ballast in ships. Charles Darwin noted in 1835 when the ship *Beagle* put in near the nitrate works at Iquique that "The nitrate of soda was now selling at the ship's side at 14/- per 100 pounds: the chief expense is its transport to the sea-coast." Transportation of the caliche to ports for shipment was always its greatest expense; the chief competitive advantage of guano was that it could be loaded directly over the islands' cliffs into the holds of waiting ships. But as the price of guano rose in the mid-nineteenth century, so did the cost effectiveness of transporting caliche to the coast for international shipment.

The great demand for guano, and later caliche, created a dire labor shortage in the guano-mining regions. Digging and processing the guano and loading it into the holds of the anchored clipper ships was a low-technology, labor-intensive business. It was a brutal and repugnant commerce to furnish something so basic and innocuous as fertilizer for crops, and to a lesser extent saltpeter for gunpowder. The economic historian Jimmy Skaggs in his study of the guano trade, *The Great Guano Rush: Entrepreneurs and American Overseas Expansion,* noted that the major source of Peruvian guano, the Chincha Islands, located approximately 120 miles south of the Peruvian port of Callao, "were the most dreadful of them all. People in Pisco could even smell them. Tropical heat, high humidity, and virtually no rainfall made the Chinchas a hell on earth. Thick, seasonal fogs that shrouded them mostly at night metamorphosed outer layers of bird droppings into a greasy paste, which baked during the day into hard crust that only picks and shovels could penetrate."

Working and living conditions on the Chincha Islands were ghastly. The workers were fed a meager fare of maize and plantains, slept in communal reed huts, and worked nearly twenty hours each day, six days a week to meet crushing mining quotas. Men choked on the acrid chemical dust when the winds picked up. They were remorselessly prodded by pitiless overseers to chip and shovel the valuable powder from the earth and load it into wheelbarrows. The barrows were pushed across the bleak, rugged surface of the island by other workers who were chained to the handles because their hands were so cracked and bleeding that they were useless for other work such as wielding heavy tools. Along the cliffs edging the island, they dumped their loads down great canvas chutes through which the guano tumbled into the bellies of waiting transport ships, sending toxic clouds of dust billowing upward. The sailors clambered into the rigging to avoid inhaling the dust that covered everything on the islands, while the hapless indentured workers, with bandannas covering their faces, rushed below in brief fifteen-minute intervals to stow the guano in sacks before emerging gasping for fresh air and spitting up blood. Disease and sickness were common and included histoplasmosis and shigellosis from accidentally ingesting the feces, as well as a host of bewildering ailments brought on by constant exposure to the ammonia dust: respiratory illness, internal hemorrhaging, and severe gastrointestinal irritations.

Newly arrived visitors to the guano islands were unable to breathe for the ammonia stench and were blinded by the dust that burned the rims of their eyes for days afterward. It was not uncommon, according to reports of travelers who visited the guano mines, for despondent laborers to commit suicide, rather than continue their remorseless toil, by flinging themselves off cliffs, rolling themselves down the canvas guano chutes into the hold of a waiting ship, slitting their own throats, or swallowing an opium overdose. Some were weakened and died from the brutal lashings of the masters; others lived in terror of the host of feral dogs used to keep them in line. Skaggs notes that "cemeteries . . . reportedly overflowed with the rotting corpses of those who failed to survive and who had been buried in graves too shallow to protect remains from scavenging dogs, human bones consequently being liberally scattered about."

In 1854, the American traveler and journalist George Washington Peck wrote after visiting the Chincha Islands that the workers were "condemned to be diggers of guano; their labours much more severe and injurious than railroad digging; they have no liberty days, no protecting laws, no power to obtain even the pittance said to be paid to them, no proper seasons of rest. Most of them go nearly naked; none have more than enough clothing just to cover themselves; they live and work like dogs; they are constantly within reach of the thongs of hideous black drivers, the link between men and devils; there are no women among them, nothing to mitigate their hopeless toil." After his ship had put to sea Peck wrote in disgust that "the islands seem to me to be a kind of human abattoir, or slaughter-house of men; and I feel a relief in being away from them, as one feels who has escaped out of some gloomy dream." Another commentator, Royal naval officer John Moresby, wrote in 1850 that the guano trade was "a system scarcely to be paralleled for cold-blooded cruelty. . . . The blackest tale of human wickedness it has ever been my lot to unravel."

Conditions were so foul that few would willingly sign on to work in the guano mines. In order to meet the need for low-paid workers, the Peruvian government and its monopoly contractors looked to Asia to fill the gap. Thousands of Chinese coolies were shanghaied to work the great guano islands, an occupation that was little better than slavery with the respectable veneer of a contract. Although Chinese labor brokers often inserted a clause in workers' contracts prohibiting their deployment in the guano trade, once at sea the contracts were poorly enforced and protests went unheeded (building railroads, mining, and farm labor were the usual jobs officially contracted for). Once the unfortunate Chinese arrived, they were virtual prisoners, forbidden to leave for the duration of their five-year contract, and few of them ever survived that long. Historians estimate that in the mid-nineteenth century between 90,000 and 150,000 Chinese were shipped to Peru in American, British, French, Spanish, and Peruvian vessels. About 10,000 died en route, and approximately 20 percent of the survivors perished after being off-loaded to mine guano on the Chinchas Islands without their consent.

Joining the unfortunate coolies were kidnapped Polynesians and convicts from the mainland, black slaves, and army deserters. A horrible practice known as "blackbirding" also supplied thousands of workers to the guano islands. American and European vessels commanded by unscrupulous and vile masters would lure Polynesian and Melanesian men aboard with promises of great wages for contract work. The ill-fated dupes were then deposited as virtual slaves to work in forestry, fruit plantations, or mineral mines, with the particularly unfortunate going to the guano islands. Thousands were brought to Peru in this manner. In 1862 a particularly sad tale of blackbirding resulted in international outrage over the practice. Several Peruvian ships sailed to Easter Island and kidnapped virtually the entire male population, about a thousand men, including political and religious leaders, and sent them to the guano islands, where most of them died. An outcry raised by a French priest resulted in the few remaining survivors being returned to their home (where they spread newly contracted diseases amongst the women and children, wiping out most of the island's population). The traffic in Chinese laborers slowed after 1874, when the Peruvian government made it illegal to import them, and in 1876 the British Parliament passed a law enabling ships of the Royal Navy to enforce antikidnapping laws even in international waters.

With very low labor costs, the guano trade was immensely profitable for the Peruvian government and the handful of fortunate shipping agents who held the monopoly to transport and sell it. The Peruvian president earned an astronomical salary equivalent to twice the salary of the president of the United States. The nation erected extravagant public works and monumentally expensive railway lines, abolished internal taxes for its citizens, and funded a series of expensive wars with her neighbors (Bolivia in 1842 and Ecuador in 1859). In 1865, Peru also had to defend itself against an attempted invasion by Spain, launched ostensibly because a gang of Peruvians killed some Spanish citizens while they were in Chile. Spanish warships seized the Chincha Islands with a view to commandeering the guano revenue and, because of a lingering resentment over the loss of her colonies in South America a generation earlier, reclaiming political control over Peru. The Spanish Pacific squadron then established a naval blockade of the Peruvian port of Callao, and later the Chilean port of Valparaiso when the Chileans joined with Peru to form a defensive pact. It was a halfhearted attempt, and together Peru and Chile succeeded in thwarting the Spanish invasion and preserving their sovereignty—and, most vital for Peru, preserving the stream of cash from the guano reserves.

Nevertheless, after several decades of astonishing national revenue generated from the exploitation of the guano islands (nearly two-thirds of all gross proceeds going directly to the Peruvian government), mismanagement, outright corruption, and poor decisions kept Peru in debt, wholly dependent upon guano to finance hefty foreign loans. Peru's entire annual income from the

Chincha Islands was consumed to service the foreign debt. The nation was insolvent without the guano revenue. Although guano commanded much higher prices in Europe and America than other fertilizers because of its superior quality, the lion's share of the premium consisted of Peruvian government export taxes. So when the value of guano on the world market began to decline steadily in the 1860s, it spelled disaster for Peru. It was a house of cards waiting for a gust of wind.

Although the demand for fertilizer remained as high as ever, by the 1860s guano was developing a poor reputation—not because of outrage over the barbaric and inhumane treatment of the guano miners, but because Peruvian guano's market predominance was being eroded by a host of other competing "guano" products that proved to be inferior and dragged down the value of genuine guano. The guano from other sources (islands in the Caribbean and off the coast of Africa and in the Pacific) was never as potent a fertilizer as the Peruvian product because the rainfall in these regions leached the nitrate from the accumulated bird droppings. Salesmen, not surprisingly, all claimed their product was the genuine Peruvian guano to get the higher price that it commanded. Because it was difficult to distinguish between the competing products, guano became a generic term for any fertilizer made from decaying bird feces regardless of its quality. Peruvian guano was at peak production in the 1840s, but by the 1860s the highest quality deposits on the Chincha Islands were dwindling, and by the 1870s the Chincha Islands were nearly exhausted and even the Peruvians began shipping guano of inferior quality from other islands. As the quality of guano became inconsistent, its reputation and value diminished further, as did the revenue to the Peruvian government.

As guano production declined in the 1860s, there was a great upsurge in the export of caliche from the coastal mainland of Bo-

Thousands of indentured laborers were worked to death under cruel and inhumane conditions mining guano on the Chincha Islands in the nineteenth century.

livia and Chile that further depressed demand for island guano. Ironically, the high global prices for guano encouraged by Peru made exploitation of the more distant Bolivian deposits of caliche more profitable and inspired caliche extraction in these regions. Soon a local industry was sprouting in the desert. Primitive towns and mining camps dotted the region by the 1860s, and workers trickled in. Although the work of mining the caliche was never as brutally evil as the guano trade, it was dirty work carried out far from the comforts of civilization. The vast Atacama Desert was one of the most inhospitable places on earth, with no rain for washing, no local food, no nearby towns other than ramshackle mining depots. Ill-paid laborers, many native South Americans but also imported Chinese, blasted, chipped, and dug the caliche from the surface of the earth and loaded it onto mules for transport across the bleak flats to purification centers and then to port towns such as Iquique and Antofagasta for shipment.

Although caliche had been commercially available from

around the same time as guano, demand for caliche increased in the 1850s when a huge deposit of potassium chloride was discovered in Germany that could be used to transform the sodium nitrate from shipments of South American caliche into potassium nitrate in a process patented by an English chemist named F. C. Hills in 1846. The product became known as conversion saltpeter or German saltpeter and was quickly put into general use throughout Europe, and to a lesser extent in the United States, particularly for military applications during the Crimean War (1854 to 1856). One of the useful and profitable by-products of this process was iodine, which until that time was derived from kelp in a long and costly process. In 1857, Lammot Du Pont discovered a means of using caliche for gunpowder without first converting it into potassium nitrate. His method involved glazing the purified caliche with graphite by tumbling it in a great barrel for twenty-four hours. The polished graphite coating protected the caliche somewhat from absorbing moisture. Although the "soda" powder was never as effective as regular gunpowder and was used only in mining, it did allow Americans to rely less on shipments of British saltpeter from India during the Civil War.

The value of caliche increased dramatically as a result of Du Pont's conversion process and due to the expanding demand for explosives after the invention of dynamite and guncotton and blasting gelatin in the late 1860s. Although the bulk of the caliche was used for fertilizer, it was absolutely vital for the burgeoning explosives industry in Europe and America. The explosives industry underwent a rapid and astronomical increase during the 1870s and continued to expand throughout the nineteenth century, coinciding with the growth in industry and the fashion for giant terrain-altering projects. The right to license the exploitation of the caliche deposits was a tempting prize for nations seeing the wealth accruing to Peru from the guano trade, while to the Peruvian government, securing control over the caliche would stave off bankruptcy and avoid international humiliation. The struggle for the caliche nitrates led to one of the most significant military conflicts in South America, the War of the Pacific between 1879 and 1884, sometimes referred to as the Nitrate War.

The nitrate plain of western South America is a narrow band only thirty kilometers wide and about seven hundred kilometers long that lay within the vaguely defined boundaries of three nations— a sort of no-man's-land of bleak, rainless, dusty plains that comprised the northern extremity of Chile, the southern extremity of Peru, and the western extremity, and only coastal access, of Bolivia. The plain is situated between two mountain ranges, the Andes to the east and coastal hills to the west, in a region that seldom sees rain. The origin of the caliche is still debated, but various theories suggest that the caliche nitrates are derived from bird or mammal excrement at some time in the distant past, perhaps from millions of seabirds defecating along the shores of a now evaporated soda lake. Bacterial activity on the organic matter would have created the caliche in a manner similar to the way that saltpeter is formed in rich soil. The historian M. B. Donald observes in the *Annals of Science* that "most of the early explorers reported the existence of patches of guano and bird remains near the nitrate deposits." And early travel accounts refer to the caliche as guano and describe it as a yellowish spongy earth, similar in appearance to the island guano. A competing theory, however, posits that nitrate-rich waters trickle from the surrounding hills and evaporate on the plain, leaving the caliche behind. The most recent evidence points to the extraordinary aridity of the region, causing a scarcity of microorganisms and plants that used nitrogen, as the reason for the immense accumulation of the nitrates.

The legend of the discovery of the peculiar properties of the caliche is recounted from an early Spanish source by Donald: The

This is an early twentieth-century nitrate-extraction plant in the Atacama Desert. After the War of the Pacific in 1880, Chile ousted Peru and Bolivia from the region and controlled the world's only industrial-scale organic supply that could meet the escalated demand brought about by the remarkable proliferation of high explosives and the dramatic rise in fertilizer use.

first person to observe the caliche was "a native woodcutter named Negreiros, of the Pampa del Tamarugal, by his having made a fire at a certain spot, which still preserves his name, and observing that the ground thereupon began to melt and run like a stream. He hastily reported this fact to his curé at Camina, who declared it to be hell-fire and asked to be shown the spot so as to be able to deal with it. The curé took a sample of the salt and found that it was nitre. He threw the remainder into his garden, where, to his surprise, the plants now grew better than before. A British naval officer, visiting Tarapacá some time later, paid a visit to the curé and spread the news to Europe." A traveler named Don Antonio O'Brien described the caliche beds in 1765: "The top surface of the ground is a coating of spongy and nitrate-containing rock, which is called caliche and is used to make houses. . . . Below this first layer is another of harder stone and less spongy which cannot be broken without powder and this is called cangelo and also contains nitrate." Other late-eighteenth-century travelers also observed the abundance of caliche as a crust on the earth, remarking that it covered entire valleys.

So long as Peru had the profits from guano propping up her flimsy national finances, it paid no attention to the mining of caliche nitrates in a region that was arguably partially part of its legal political territory. But by the 1870s, caliche nitrates were undermining the Peruvian economy. Consumption of caliche nitrates in Britain alone more than doubled between 1868 and 1873, while guano shipments declined by half during the same period. Seeing its export revenue decline, which funded the bulk of Peruvian government operations, the government sought to establish a monopoly on caliche as it had on guano decades earlier—by declaring unworked caliche deposits to be official government property, subject to government licensing and price control. Historians Robert G. Greenhill and Rory M. Miller write in the *Journal of Latin American Studies* that "Peru's eventual response to financial crisis and declining guano sales in the 1870s was, in absence of practical revenue-raising alternatives such as direct taxation, to expropriate the nitrate industry." They planned to control and manipulate the production and sale of caliche to prevent competition with guano.

Many of the same companies were involved in both the guano and caliche nitrate business. Greenhill and Miller note that "it was ironic that the profits which west coast houses derived earlier from the guano trade would now finance a rival fertilizer, nitrate." The commercial infrastructure, capital, and shipping network established to exploit guano were easily and quickly adapted to the expanding caliche industry. William Gibbs, for example, one of the principals of the South American–based British export firm Anthony Gibbs and Son, built an astronomical fortune for himself by astutely positioning his firm as the monopoly agent for Peruvian guano even while heavily investing in the caliche nitrate industry. In the late 1860s, he built a sprawling mansion on an estate in the English countryside west of London that Britain's National Trust describes as "a full-blooded Gothic Revival extravaganza." Moreover, the caliche industry differed from the guano trade in that it was not purely an enterprise of extraction but rather required capital investment and organization to produce a product that had international value. Most of the large-scale and successful firms that were later involved in the extraction and purification of caliche were, although nominally Chilean companies, entirely backed by foreign capital. When Peru nationalized the caliche deposits in 1875, it was an indirect blow to established Chilean and European economic interests.

While rousing the ire of established businesses, Peru's nationalization of the caliche deposits failed to arrest the decline in guano prices and produced little economic benefit—caliche nitrates were also found in the Bolivian and Chilean portions of the Atacama Desert. While Chile's deposits remained mostly uneconomical to extract because of their distance from the coast, in Bolivia the distance was not so great, and low Bolivian taxes helped to defray this cost, stimulating a burgeoning industry that was in direct competition with Peru's government monopoly. The southern Peruvian province of Tarapacá was geographically separate from the remainder of the country and was entirely dependent on exports from Chile to feed and supply the miners, who were themselves mostly foreigners. There was little loyalty to the country of Peru, and little culturally that tied the region to the nation to which it nominally belonged. Peru's political grip on the valuable territory was tenuous, making it a prime target for annexation.

While Peru sought stability and regulation of the entire nitrate plain and a cartel-like control over the price of caliche, Chile sought to expand her participation in the industry. A brief period of oversupply caused by the exploitation of the Bolivian nitrate deposits in the 1870s threatened the price stability that the Peruvian government depended upon to meet financial obligations. Fear and greed being powerful motivators, the astronomically valuable dust led the three nations inexorably toward war. The actual pretense for the conflict was that Bolivia had increased the tax on a Chilean firm (financed by British capital, including Gibbs) near the Bolivian port of Antofagasta, contravening a treaty between the two nations that guaranteed a low tax for Chilean companies extracting caliche nitrates in exchange for Chile's acknowledging Bolivian political claims to the sparsely populated region. Perceiving a threat to national interests, and desiring to control the caliche nitrate industry, a force of five hundred Chilean troops seized Antofagasta on February 14, 1879. Peru was dragged into the war through pride and a secret defensive alliance with Bolivia. The war was a series of Chilean victories against inferior Peruvian and Bolivian forces. By the end of the first year of the war, Chile had not only claimed the Bolivian province of Atacamá and the Peruvian province of Tarapacá, but had established naval supremacy by capturing a vital Peruvian ironclad steam warship called the *Huascar*. Chile then blockaded all foreign shipments of guano and caliche, denying Peru its vital foreign income. After a series of further naval and land victories, Chilean troops marched on the Peruvian capital Lima in 1881. Although Peru and Bolivia struggled on for a few years longer, they were both utterly defeated and sued for peace in October 1883 and April 1884 respectively. Both ceded to Chile their provinces in the caliche nitrate plain, significantly extending the northern reach of Chilean territory. Bolivia, in addition to being denied the wealth of the caliche, lost her only coastal territory and henceforth had no access to ports for shipping. It remains a contentious issue to this day; some Bolivian maps still show the disputed coastline as an occupied territory. Meanwhile, the official Bolivian navy sails the Bolivian portion of Lake Titicaca awaiting its triumphant return to the Pacific.

The outcome of the war was that Chile ended up with the entire caliche nitrate fields, while Peru retained a monopoly over the dwindling island guano supplies. In the following decades, the Chilean nitrate industry continued to expand (owned by foreign capital, employing menial local laborers to do the backbreaking dirty work in the desert, the taxes covering most national government revenue requirements while the profit fled the country). The extraction process became ever more mechanized and efficient, with steam shovels to dig and scrape away the earth, and steam locomotives to transport caliche to the coast. Refined sodium nitrate was loaded into ships at ports Antofagasta and Iquique and transported directly to manufacturing plants like the giant British Dynamite Factory in Scotland to fuel the insatiable and expanding demand for explosives and fertilizer. The relatively low price and ease of shipping Chilean caliche nitrates undermined the few remaining local saltpeter industries in Europe and even India saw a steep decline in saltpeter production for export. Chile was left controlling virtually the entire global supply of industrial-scale commercial nitrates on the cusp of the world's greatest increase in demand.

Already in great demand as a fertilizer, with the astronomical increase in the use of dynamite throughout the late nineteenth century, caliche became a vital raw material of war and agriculture. "It may, therefore, be safely asserted," claimed the historian C. E. Munroe in *The Nitrogen Question from the Military Standpoint*, "that but for the discovery and exploitation of the nitrate fields of Chile the explosives industry, as it is known today, would have been impossible, and the developments in mining and transportation which have characterized the last half of the nineteenth century would not have been made." It is not surprising that the first major naval battle of the First World War was fought over Chilean nitrates.

In 1896, just as the Chilean nitrate industry was about to have another banner year, producing an astonishing 1,158,000 tons of nitrate for export, nine times the quantity produced before the start of the War of the Pacific, the man who had done more than any other to generate the nonagricultural demand for the product, Alfred Nobel, was suffering a worsening of the heart problems that had plagued him for years, which prompted him to draft a last will and testament in Paris. Over the course of his three decades as the world's preeminent explosive researcher and manufacturer, he had developed unusual opinions about the future use of the financial rewards from his life's work. His unorthodox will would have no impact on the Chilean nitrate industry, of course, or upon his dozens of explosives factories that regularly received huge shipments of purified caliche to make their products, but it was to have an impact on the world.

The Profits of Dynamite
A Gift to Science and Civilization

*"Since the actual transfers to and from the Consulate involved
certain risks of hold-ups and robberies, special precautions were
taken and exercised to avoid attention. . . . after the securities
had been packed in a suitcase, we took an ordinary horsecab to
the Consulate General. With a loaded revolver in my hand, I sat
in the cab prepared to defend the suitcase in case a collision with
another carriage had been arranged."*
—Ragnar Sohlman, ca. 1890s

San Remo on the Italian Riviera was a peaceful, secluded spot removed from the main European thoroughfares. It was the opposite of a grand city like Paris, which was the intellectual and geographical crossroads of Europe. San Remo was not the center of anything. People went there to relax and unwind, to disengage from the commercial and artistic currents of their lives. The weather was pleasant and sunny, the climate clement and mild. Alfred Nobel did not like it. He preferred Paris. Although San Remo was beneficial for his health, it did not suit his temperament, and he found himself frequently away on business, on the road to Scotland, to Sweden, to Germany and France. Because of San Remo's remoteness, all chemicals and lab equipment had to be ordered from Germany at great delay, and educated assistants were not available in the local population. Neighbors complained of the explosions from his laboratory and the launching of rockets and firing of ordnance off the giant pier that jutted into the water from the grounds of his sprawling estate. It was not an ideal location for many reasons.

But Nobel, who had turned sixty just before he was driven from Paris in 1891, was growing worn-out from his ceaseless travels and constant work. "I am sick of the explosives trade," he wrote to an assistant, "wherein one continually stumbles over accidents, restrictive regulations, red tape, pedants, knavery, and other nuisances. I long for quiet and wish to devote my time to scientific research, which is impossible when every day brings worry." His nitroglycerin headaches had also grown progressively worse over the years. At first he would get them only when he was working in his lab, inhaling the fumes of the nitroglycerin and solvents with which he frequently experimented, but by the 1890s he was routinely afflicted with "repeated visits from the spirits of Niflheim." Niflheim refers to the land of mists, dark, and cold that houses the realm of the dead in Norse mythology. He suffered from bronchitis every winter. French medical experts also diagnosed angina pectoris and prescribed nitroglycerin as a treatment to counteract what Nobel called "rheumatic devils paying a visit to the heart muscles or thereabouts." In a highly diluted form nitroglycerin had been found to alleviate the symptoms by dilating blood vessels and increasing oxygen supply to the heart. "It sounds like the irony of fate," he wrote to his assistant Ragnar Sohlman, "that I should be ordered to take nitroglycerin internally. They call it trinktin so as not to scare pharmacists and public." Doctors ordered him to slow down, rest from work, but this was something he could not do. It was not in his character, and he kept up his grueling travel. Although he had resigned from all the directorships of his numerous corporations and trusts, he kept up his lab work late into the night.

Nobel was a born inventive genius, and his fertile mind continued to come up with new ideas and products. He couldn't rest even if he had wanted to. During his lifetime he amassed over 355 patents in different countries, which Eric Bergengren describes with understatement as "an impressive number to have originated in a single brain." His ideas included tools for aerial photography and blood transfusions, prototype artificial rubber and artificial silk, as well as numerous variations on his original explosives patents and improvements to firearms. As purely technical problems the work fascinated him but, paradoxically, by this time in his life he had developed a strong aversion to war and the practical application of his martial discoveries and improvements. The research that occupied him in the last years of his life was seemingly at odds with his personal philosophy. "For my part," he wrote, "I wish all guns with their belongings and everything could be sent to hell, which is the proper place for their exhibition and use." Nevertheless, Nobel continued to work toward more reliable fuses, silent firing guns, more stable and predictable projectiles, better-sealed firing chambers, and the like.

After a year or so, Nobel found San Remo too remote, sedate, and unsuited to his lifestyle, and he cast his eyes about the map of Europe, looking for a suitable spot where he could pursue his experiments unhindered by authorities and yet not so remote as to be an intellectual backwater. In his wry, sardonic manner, he dismissed England, not because of the cold and damp, but because "In England conservatism is too flourishing for counsel to accept anything which has no antediluvian sanction." He likewise dismissed France, not only because he had been treated shabbily and unceremoniously ousted, but because he felt the French were too haughty: "All Frenchmen," he wrote, "are under the blissful impression that the brain is a *French* organ."

Germany, with its swaggering and aggressive military culture, made him feel uneasy. He began to take a greater interest in the land of his birth. Although he had not lived in Sweden for three decades, since the 1860s, he had always visited his mother each year, and he mused about setting up a home in his native country. In 1893 he was awarded an honorary doctorate in philosophy from the University of Uppsala and this solidified his resolve. Although he afterward quipped: "since the creatures have made me a Doctor of Philosophy, I have become almost more of a philosopher than before," he was nevertheless touched by the honor and enjoyed being referred to as Dr. Nobel. In 1894, at Bofors, about fifty miles from Stockholm and only a few miles from his brother Robert's retirement estate, he bought a giant ironworks and munitions manufacturing plant that had fallen on hard times and infused it with new capital. He renovated the manor house and built a grand laboratory for himself and his numerous assistants to continue his work.

By this time his health was failing, and he knew it. With his heart giving him ever more trouble, his doctors in France advised him to order his affairs. He had earlier written, "I have no memories to cheer me, no pleasant illusions of the future to comfort me, or about myself to satisfy my vanity. I have no family to furnish the only kind of survival that concerns us, no friends for my affections or enemies for my malice." And on another occasion he wrote that, "For the last nine days I have been ill and have had to stay indoors with no other company than a paid footman; no one inquires about me. . . . My heart has become as heavy as lead. When at age fifty-four, one is left alone in the world, and a paid servant is the only person who has so far showed one the most kindness, then come heavy thoughts, heavier than most people can imagine." Nobel had been thinking about his legacy for several years, ever since the death of his older brother Ludvig. In 1888, Nobel's obituary appeared erroneously in a Paris newspaper. It described his discovery of dynamite and detailed his rise to riches as a purveyor of explosives and weapons, painting him as an immoral man who profited from human conflict and misery, providing the means for others to maim and kill each other. Nobel was shocked when he read it, for it was his older brother Ludvig who had died, and the reporter obviously had confused the two of them. It prompted Nobel to ponder his life, however, and to evaluate his contribution to the world. He also began considering, having no direct descendents, what to do with the proceeds of a lifetime of accumulated wealth, proceeds from the invention, manufacture, and sale of perhaps a million tons of explosives—not all of it used in the service of mankind. At the time his fortune hovered somewhere in the range of 33 million Swedish kronor (which was equivalent to about 2 million pounds sterling or 8 million U.S. dollars). Direct comparisons to today's currency are difficult and ultimately misleading, but it was truly a mighty sum.

Yet even while pondering his moral and ethical obligations, Nobel continued not only to manufacture and sell explosives and smokeless powder to the military, but pursued his research in these fields, hiring numerous assistants to increase the pace of his experiments. Nobel had a brilliant practical mind and easily justified his business decisions by pointing out that he was merely producing tools that others could then use as they wanted, for practical benefit or terrible evil according to their inclination. It was not his responsibility, he claimed, and if he wasn't producing the explosives and weapons, others would surely fill the vacuum with similar products. Logically, he was usually on solid ground, but Nobel was a man with a conscience and his moral responsibility began to weigh heavily on his mind. His work was his life, yet in his final years it became ever more clear that work alone was not completely fulfilling. "I have great things to think about," he wrote a year before his death, "at least one great thing—the passing from light to darkness, from life into the eternal unknown or, as Spencer calls it, the unknowable."

In July of 1896 Nobel's brother Robert unexpectedly died on his estate not far from Bofors. Alfred, the least robust member of his family, had outlived all his siblings and both parents, and he knew his own demise was imminent. When he departed Sweden in the fall for the milder climate of San Remo, he stopped in Paris to visit a heart specialist and was advised to cease work and travel and begin to arrange his affairs. Nevertheless, he continued to work in his lab in Italy and looked forward to returning to Sweden the following spring to continue his experiments with ballistics and guns. But it was not to be. Early one morning he was found slumped in a chair, having suffered a stroke. In the final few hours of his life, he began babbling in a language none of his servants could understand—Swedish. He managed to scrawl out the word "telegram," and news was quickly sent to some family members and to his assistant Ragnar Sohlman, but no one could reach Italy before his death early on December 10, 1896. One of his biographers, Herta E. Pauli, has written that his "end came as quietly as he tried to live—a sad and perfect closing. . . . He went to his death alone—a shy, inconspicuous, colorless man. His neighbors did not know he was dying. No one was there. . . ." His body was taken to Sweden for cremation. The last letter he wrote, unsent on the desk where he suffered his stroke, was to an assistant stating that although his health was bad he would continue working in his lab when he felt stronger.

When he died, Nobel had over ninety-three factories producing 66,500 tons of dynamite or dynamite-related explosives each year. He owned shares in factories and businesses spread over at least nine countries. He had six large mansions in five European countries. He owned a great deal of real estate, intellectual property, and other assets, and many were expecting his will to provide a detailed breakdown of his estate and precise instructions for its distribution. The will, however, proved to be a complete shock, not only to his extended family but to all of Europe. The previous November, soon after the dispiriting news of the loss of the cordite lawsuit in England, Nobel had written by hand in Swedish a concise four-page document while he was in Paris and had it witnessed by four members of the Swedish Club. He deposited his will in a vault in Stockholm, where it lay for a year like a time bomb waiting to explode. His dislike of lawyers, exacerbated in the aftermath of the cordite lawsuit, which vexed him and insulted his sense of fairness and justice, led him to draft his will without any recourse to legal advice. As a result, it was a bewilderingly vague and insubstantial set of instructions to dispose of a vast and diverse fortune.

After a brief preamble and a list of some small personal bequests, Nobel launched into a confusing and imprecise description of what was to happen with the bulk of his fortune:

> The whole of my remaining realizable estate shall be dealt with in the following way: The capital shall be invested by my executors in safe securities and shall constitute a fund, the interest on which shall be annually distributed in the form of prizes to those who, during the preceding year, shall have conferred the greatest benefit on mankind. The said interest shall be divided into five equal parts, which shall be apportioned as follows: one part to the person who shall have made the most important discovery or invention within the field of physics; one part to the person who shall have made the most important chemical discovery or improvement; one part to the person who shall have made the most important discovery within the domain of physiology or medicine; one part to the person who shall have produced in the field of literature the most outstanding work of an idealistic tendency; and one part to the person who shall have done the most or the best work to promote fraternity between nations, for the abolition or reduction of standing armies and for the holding and promotion of peace congresses. . . . It is my express wish that in awarding the prizes no consideration whatever shall be given to the nationality of the candidates, but the most worthy shall receive the prize, whether he be a Scandinavian or not.

It was an unprecedented last will and testament. From his scraggly handwritten document has emerged one of the richest and most prestigious scientific prize agencies in the world, distributing millions of dollars annually for over a century so far. Even one fifth of the interest of such a vast fortune was enough to enrich an individual, and at the turn of the nineteenth century, it was calculated that the amount of each prize would be equivalent to thirty times the annual salary of a university professor or one hundred times the annual salary of a skilled construction worker.

The five prizes stipulated in the will reflected Nobel's personal philosophy—that discoveries, inventions, and ideas would produce great social benefits. Nobel himself was always an avid reader of literature and an amateur novelist, playwright, and poet. Physics and chemistry reflect Nobel's interests and scientific pursuits while medicine was of obvious potential benefit to mankind. It was Nobel's belief that every discovery, every new invention that expanded and improved human knowledge, directly contributed to an improvement in society, a betterment of conditions and a decline in evils. A rumor at the time was that mathematics was not included because Nobel personally disliked a prominent Swedish mathematician named Gösta Mittag-Leffler—or that he stole a woman from Nobel. It has been dismissed as untrue by Nobel's biographers. Mathematics was not included because Nobel found it too theoretical and not of direct practical benefit.

Perhaps most surprising was his inclusion of peace as one of the five prizes. Although the sentiment that people should refrain from murdering each other and destroying civilization was one that Nobel professed for most of his life (a seemingly incongruous sentiment inasmuch as conflict was the source of much of his wealth) the inclusion of a peace prize, alongside the more tangible science prizes, was likely attributable to Nobel's friendship in the final years of his life with the Countess Bertha Kinsky. She may have been, in fact, partly responsible for Nobel devoting his life's accumulation of wealth to a socially progressive end in the first place. Although he had written two earlier wills, one in 1889 and one in 1893, each with provisions for some of his fortune to go to prizes, Nobel's final will, written after several philosophical discussions and years of correspondence with Kinsky, left nearly his entire fortune for this purpose and cut out his family.

Nobel first met Kinsky in 1876 when he posted a cryptic advertisement in a Vienna newspaper. The advertisement proclaimed that "a wealthy, highly educated elderly gentleman" in Paris was searching for a secretary of "a mature age, with a knowledge of several languages." The thirty-three-year-old woman who applied was from an old aristocratic Austrian family that had fallen on financial difficulty. She had been employed as a governess to another wealthy family but had reluctantly departed when her budding affair with the family's eldest son was discovered. Having nothing to fall back on, she applied for the unusual posting in Paris, where she developed an immediate rapport with the far from elderly gentleman (Nobel was forty-three at the time). Within a week, however, while Nobel was in Sweden, she returned to Vienna and eloped with her lover to the Caucasus. Kinsky and Nobel remained in touch through letters for the next decade before she returned to Vienna. A decade and a half later, when she wrote her triumphant novel *Lay Down Your Arms*, a passionate defense of peace and disarmament which was translated into numerous languages and made her famous throughout Europe, their relationship took an entirely new turn. After the book's publication, Nobel wrote to congratulate her, referring to her as an "amazon who so valiantly wages war against war." Through letters and in person on several occasions, she sought to persuade the Dynamite King to embrace her—radical for the time—doctrine of peace and arms reduction in the growing European peace movement.

Although Nobel was not a warmonger, his views were more pragmatic than idealistic. Kinsky's impassioned pleadings for immediate disarmament and her belief that it would come soon if

Alfred Nobel's last will and testament was surely one of the most unusual and surprising legal documents of the era.

enough protests and peace congresses were organized, were somewhat naïve and failed to account for the violence in human nature or the extent to which conflict and war are an entrenched appendage of culture with proud martial traditions dating back to the dawn of humanity. War, whether out of aggression, for defense, or to right past grievances, has proved most resilient as a dispute-settling mechanism and wasn't about to fade away in a few years because of a growing number of peace congresses—and hasn't in the century since Nobel died. Indeed, the twentieth century has been one of the most bloody, with two major world wars, and the twenty-first has not begun any differently. The idea for an international forum for the arbitration of conflicts, however, one of the prime objectives of the peace movement, was a sound concept, and Kinsky's passion and earnest appeals for moral and financial support did eventually have an impact on the skeptical and cautious explosives magnate.

Nobel's antiwar sentiments had been established in his youth, and were not merely the whimsical musings of a lonely man in his closing years. But his early opinions, as much as they were opposed to murder and destruction, amounted to a pragmatic excuse through which he could absolve himself from moral responsibility for the uses to which his inventions were being put. Kinsky reported in 1876 a conversation she had with him during her brief employment as his secretary. "I wish I could produce a substance or machine of such frightful efficacy for wholesale devastation that wars should thereby become altogether impossible," he claimed. Nobel believed that war would kill itself when the potential for destruction became too great for either side to risk a battle. And this remained his avowed philosophy until a few years before his death, despite never having come across any evidence to support it (although the early end of the Franco-Prussian War with dynamite initially bolstered his opinion). In 1892 in Zurich, Kinsky related their conversation as they cruised the lake on a boat in the evening:

"Perhaps dynamite factories are even more profitable than silk mills—and less innocent," she claimed. To which he responded, "On the day when two army corps will be able to annihilate each other in one second, all civilized nations will recoil from war in horror and disband their forces." But as one of Nobel's biographers, Herta E. Pauli, astutely observed, "what then if an uncivilized one appears? What if this one turns to the civilized nations: Be reasonable—why should we wipe both our cities off the map—all I want is a little corridor, my last territorial demand—why let our women and children writhe in agony when I drop gas—do you want to die for Danzig?" She points out that the "theory, of increasing the horror of war to deter men from waging it, is one of the most colossal fallacies in the history of human thought." It was also a philosophy in which Nobel had lost faith by the 1890s. The idea that more destructive weapons would result in a more peaceful world does have some practical truth to it, in the sense that a country is less likely to assault a strong neighbor over a minor disagreement; however, it does not hold up as the moral justification for the creation of those weapons, particularly if one plans to sell them to the highest bidder rather than to give them away.

It almost seems as though Nobel's philosophy was an elaborate self-deception, allowing him the illusion that the field of his technical fascination, and his life's work, was purely honorable and productive. In addition to being a persistent technical genius, Nobel was an introspective, poetic, and somewhat gentle soul. He did not want to believe that the source of all his wealth, success, prestige, and honor was merely greater power for destruction and better methods for killing. As the years passed, and with the lessening of his driving ambition to succeed, he became increasingly uneasy with the uses to which his inventions were being put, even while being unable to separate the emotional belief that it was wrong from the logical belief that he merely supplied the tools for others to use as they pleased; he had a conscience, and as he aged and began thinking of his own death, he wanted more than to be known as an eccentric multimillionaire who made his fortune as a merchant of death.

Although the extent to which Kinsky influenced Nobel's decisions regarding the distribution of his estate has caused much speculation, ultimately it is impossible to know. Pauli has commented that "the precise amount of credit due her for his active peace work has been much discussed and violently debated. It is futile speculation; no one but they themselves knew—and they did not tell." In any case, it was a combination of his own introspection and concern for his legacy combined with Kinsky's influence as a charter member of the late-nineteenth century disarmament movement that led Nobel to will his estate to support international prizes in the practical scientific fields, and to the peace prize specifically. In drafting his will, Nobel may have believed that using a portion of his fortune to oppose the conditions that gave rise to it would result in a primitive balance, cleansing the moral taint from his scientific and industrial activities, which he knew were stained in blood—beginning with the death of his youngest brother Oscar-Emil, leading to hundreds of others as a result of his premature and ill-conceived venture into nitroglycerin production, and culminating in the development of his most powerful explosive, blasting gelatin, and smokeless powder, ballistite, for primarily military use.

Irreconcilable contradictions were the most distinguishing characteristic of Nobel's personality. He was a poet and writer as well as the archetype of the nineteenth-century international capitalist. He was a man so driven to succeed that it damaged his health and ruined his chances for a family of his own, yet he took little pleasure in the trappings of wealth. He was a man devoted to his family who labored and gambled to help establish the fortunes of his brothers but who left little of his own fortune to any family member when he died. He was a lonely man who longed for social companionship, yet shied from society and spent most of his days alone in his laboratory. He was a man who disliked the inconveniences of travel, yet spent most of his days traveling for business. And he was a hard-nosed merchant who fought to sell armaments and explosives to both sides in a conflict, even while harboring a deep-rooted abhorrence of violence and war. His last will and testament merely reflected the polarized inconsistencies that he wrestled with throughout his life, the irreconcilable contradictions between his mind and his heart. The moral struggles and the internal conflicts that plagued him in life would be resolved in this final document.

When Ragnar Sohlman and Rudolf Lilljequist obtained Nobel's will from the bank vault in Stockholm and read it, they must have despaired of ever fulfilling the requests. Although Sohlman had worked for Nobel for three years as a personal assistant at San Remo and at Bofors and was described by Nobel as "one of my few most favorite people," he was only twenty-six years old and had no legal experience. Lilljequist, although forty years old, had only met Nobel twice and also had no legal experience. He likely chose these individuals because of their integrity and honesty, traits he found lacking in other associates and business partners. Sohlman and Lilljequist were also Swedes who had lived abroad like himself. Nobel had told a friend once that he wanted his estate administered by Swedes because "it was in Sweden that he had met the greatest proportion of honest men." Since he was leaving nothing to his relatives apart from the small legal stipend required by Swedish law, Nobel chose to leave his family out of the administration of his estate. Of his two administrators, Sohlman in particular shouldered an immense burden when he accepted these responsibilities. These estate administration duties would consume him for the better part of five years, lead him into struggles with governments, Nobel's powerful family, and the scientific institutions that Nobel had selected to award his prizes. The Nobel Foundation might never have come into existence, owing to the great problems with the will's legality and cavalier format. Although Nobel despised lawyers and didn't want them meddling with his personal estate, particularly after the disheartening British ruling in the cordite case, his decision not to have a legal expert draw up his will very nearly doomed his wishes. Despite the obvious intent in the will, and its inherent simplicity and brevity, it was a tangled legal nightmare of questionable legality and open to interpretation. The first and most obvious question that arose was that of Nobel's domicile, for the will had to be interpreted and the estate administered according to the laws of his country of domicile. Although it was Nobel's stated opinion that he wished his will to be administered in Sweden, his country of residence was of greater legal significance—and for a wealthy gypsy like Nobel, residency was not immediately obvious. Nobel was not a resident of any nation in the usual sense, and although he had written his will in Swedish, named two Swedish executors, deposited the will in Stockholm, and asked Swedish scientific institutions to decide the prizewinners, a strong argument could made for his residency in other countries as well. He had lived in France for eighteen years and he still maintained his mansion in Paris, yet he also had large houses in Germany and Scotland. At the time of his death, he spent his winters at his estate in San Remo and his summers in Sweden. He had signed the will and had it witnessed in Paris before taking it to Stockholm. If he was declared a French resident, the death duties would have consumed a great portion of the estate. In addition, under French law, the will would probably have been declared invalid, along with the prizes, and the remaining assets after taxes would have been divided among his family.

The conflict over the disposition of such a vast sum of money resulted in an all too typical tawdry affair in which branches of the family quarreled and sought to have the will broken. Bergengren writes: "There was indeed an intricate and seething conglomeration of arrangements, viewpoints, attempted coups, and protracted lawsuits, into which financial, scientific, and legal experts of several countries were drawn; the press, with its sharp comments for and against every step taken, did not make the situation any easier for the executors, who were valiantly fighting as proposers, persuaders, and mediators." The Swedish branch of Nobel's family, the descendents of his brother Robert, secretly sought to have Nobel declared a resident of France so that the will would be found void and they would be entitled to a greater share of the proceeds. In the end, Sohlman and Lilljequist, with the help of legal counsel Carl Lindhagen, and a host of other legal specialists throughout Europe, worked out the many difficulties and resolved the conflicts. They convinced the head of the Russian branch of the family, Ludvig's son Emanuel, to declare in favor of the will by agreeing to sell him Alfred's shares of the Nobel Brothers Naphtha Company in Russia for a good price. With Emanuel and the Russian branch of the family refusing to contest the will, the momentum from the Swedish Nobels petered out and Alfred Nobel was declared a resident of Sweden. The battle continued, however, even after Sweden became the legal jurisdiction and dragged on for several years, until the Swedish heirs were offered an undisclosed sum and finally dropped their lawsuit.

The second great hurdle to overcome was that the foundation which was to administer the funds and adjudicate the prizes did not yet exist, and the academic institutions designated to select the prizewinners had never been forewarned of the crucial role Nobel had imagined for them. How could money be bequeathed to a nonexistent foundation? It took several years for Sohlman and Lilljequist to broker the agreements between the institutions and establish exactly how each institution was to be compensated for its expenses and time. The responsibility for choosing the most significant international writers and scientists was a daunting undertaking. Awarding international prizes in science and literature required highly informed specialists capable of understanding the bewilderingly broad and complex fields of study, and the ability to appreciate the most significant advances and discoveries. At the time, Sweden had a population of only five million, with few internationally renowned writers or scientists. Neither the Swedish Academy, responsible for deciding the prize in literature, the Royal Swedish Academy of Sciences, responsible for the chemistry and physics prizes, nor the Karolinska Institute, which was responsible for the prize in medicine, had any qualifications or background to assume such lofty and prestigious global responsibilities. Even more problematic was Nobel's assignment of responsibility for the peace prize to the Norwegian parliament, an action that aroused patriotic ire in Sweden. Even the Swedish press called for abolition of the will over this fact alone. Burton Feldman writes in his book *The Nobel Prize:* "In 1900, how competent were the Swedish academies and Norwegian parliament to award Nobel's prizes? The brief answer must be: not very." Nevertheless, they did eventually rise to the occasion.

Collecting the disparate assets of Nobel's far-flung financial empire was an arduous task in itself. All the houses had to be sold, the shares converted to cash, and the funds transported to Sweden. The legislation governing the disposition of real estate and stocks was different in each of the nine countries containing portions of Nobel's fortune. In an age of bearer bonds and no electronic money transfers, or even reliable international agreements, the wealth all had to be collected and transported physically. Sohlman describes a somewhat amusing scene in Paris of his hauling mounds of negotiable securities from several Paris banks and Nobel's house in briefcases and skulking across town to the Swedish consul general's office. The only insurance they could obtain had a daily limit that required them to make dozens of journeys hauling the equivalent of millions of dollars. "Since the actual transfers to and from the Consulate involved certain risks of hold-ups and robberies, special precautions were taken and exercised to avoid attention. . . . after the securities had been packed in a suitcase, we took an ordinary horsecab to the Consulate General. With a loaded revolver in my hand, I sat in the cab prepared to defend the suitcase." Sohlman then hauled and guarded the suitcases of securities from the consul general's office and en route to the railway station and eventually all the way back to Sweden.

Just as the years of work were finally nearing completion—the disinherited heirs placated, the academic institutions brought onside, the vast fortune in assets liquidated and carried to Sweden and invested in safe securities—a shocking revelation came to light regarding a hitherto unknown claimant on Nobel's estate: Sophie Hess, Nobel's companion for eighteen years, threatened to reveal nearly two decades' worth of potentially embarrassing love letters if she was not granted a lifetime annuity. In the end she was given an undisclosed sum in exchange for all the letters, which the Nobel Foundation promptly hid for decades, even from biographers, until the 1950s. On June 27, 1900, the king of Sweden, Otto II, signed into existence the Nobel Foundation, which took over responsibility for the funds from Nobel's executors. One year and a half later, on December 10, 1901, the fifth anniversary of Nobel's death, the first prizes were finally awarded.

The creation of the Nobel Foundation transferred and channeled the amassed profits from the greatest international explosives and armaments empire of the era in an entirely new direction, toward the betterment of society and the promotion of civil discourse between nations. Right from the start, the awards were attended with controversy and disagreement. One of the early recipients of a Nobel Prize in Chemistry, the first award after the prizes were suspended for several years during the First

World War, was a brilliant and enigmatic German chemist named Fritz Haber. Haber's award-winning discovery was a natural extension of Nobel's interest and research into explosives—the field of research that had generated the prize money in the first place—but the uses to which it was put would have appalled and horrified Nobel had he lived to witness it. The four years of the First World War, 1914 to 1918, had created a great demand for explosives and their vital raw material, which again focussed international attention on the remote coast of Chile.

The Battle of the Falklands
The Struggle for the Global Nitrate Supply

To utilize fully our existing high explosive plants it is neces-
sary that we should ship from Chile approximately 788,000
tons of nitrates; at present the tonnage for only 600,000 has
previously been agreed upon. New and very serious demands
for TNT are also being made by the Admiralty for mines.

—Winston Churchill, 1917

In the fading sun of a late spring evening on November 1, 1914, a German cruiser squadron unexpectedly encountered a British squadron of similar size in the turbulent waters off the coast of Chile near the small town of Coronel. After jockeying for advantageous position, the steel hulks closed and joined battle. A painting of the engagement, one of only a handful of decisive naval battles in the First World War, shows the German armored cruiser *Gneisenau* plunging through gray wind-whipped waves. Froth spatters over the desk of the ship as it rolls in the heavy sea, and the setting sun illuminates the sky and clouds in an orange glow. Eruptions of shot speckle the water surrounding it while flames lick along the compartments above the water line trailing oily smoke, the evidence of several direct hits. In the background of the painting, traces of the other ships, three British and four German, can be seen blasting away at each other with their great guns. The battle wore on for several hours of deafening cannon fire and rumbling explosions, until the approaching darkness reduced visibility and silenced the guns. Not yet three months into the war, it was the first significant naval battle of the conflict.

At the outset of the war in August, the German Far East Fleet, the most powerful and dangerous German fleet outside Europe, was scattered throughout the South Pacific. Under the command of Admiral Maximilian Graf von Spee, a stern fifty-three-year-old south German aristocrat, the ships rendezvoused soon after the war began and consolidated into a seven-ship squadron. As the historian Ronald H. Spector writes in *The Great War:* "Von Spee's squadron had been a major headache for the British since the first days of the First World War. Sightings and rumors of sightings of his ships, and fears as to what they might do, had caused panic throughout the Far East and South Pacific. Merchant shipping was thrown into confusion or halted. Troop convoys from Australia and New Zealand were delayed or rerouted. More than twenty British and French cruisers, the entire Royal Australian and New Zealand navies, and a substantial portion of the Japanese navy (which had entered the war on the Allies' side in mid-August) were occupied in searching for von Spee." After sending one cruiser to the Indian Ocean to prey on shipping and another to the South Atlantic, von Spee rendezvoused the other five ships, the armored cruisers *Scharnhorst* and *Gneisenau* and the light cruisers *Nürnberg, Dresden, Leipzig,* and half a dozen attending steamers at Easter Island on October 14, 1914.

Well provisioned, well coaled, and under a bold and competent commander, the fleet set off east for South America. There was no immediate pressure from a superior British fleet to worry von Spee, so he initially planned a campaign of raiding and assaulting commercial shipping, in particular the indispensable nitrate freighters bound for the Panama Canal, and eventually for Europe, where they would provide the raw material for English and French armaments that were in desperately short supply. It had been over half a century since Chilean nitrates had flooded the European and American markets for agriculture, civil engineering, and, most important in 1914, the military. By the start of the war nitrates were more valuable than ever before as the crucial ingredient of all armaments—from smokeless powder in cannons and rifles to mortars, mines, and all other explosives. "It was not a purposeless adventure that brought Admiral von Spee's fleet across the Pacific," wrote Grosvenor Clarkson, the director of the U.S. Council of National Defense during the war. "To strike at the source of the Allies' nitrate supply was to paralyze the armies in France. The destruction of a nitrate carrier was a greater blow to the Allies than the loss of a battleship."

The closest British squadron to the west coast of South America were five ships under the command of Sir Christopher Craddock, a distinguished and decorated fifty-two-year-old Yorkshireman who entered the Royal Navy at age thirteen as a midshipman and slowly rose in rank to vice admiral. When the British Admiralty intercepted a wireless transmission of von Spee's probable destination of Chile to raid commerce, Craddock was ordered, in a series of confusing and vague wireless transmissions, from his posting in the South Atlantic to round Cape Horn and cruise the coast of Chile as far as Valparaiso searching for the elusive German squadron. He was also ordered to attack German commercial shipping and protect British commercial shipping—by its very presence, von Spee's squadron had put a halt to the activities of the British merchant nitrate fleet. Craddock's squadron consisted of his flagship the *Good Hope,* the aging and decrepit armored cruiser *Monmouth,* the poweful light cruiser *Glasgow,* and the armed merchant ship *Otranto.* Three hundred kilometers astern, convoying the fleet's colliers, was the ancient battleship *Canopus,* far slower, armed with inferior guns, and manned by an untrained crew of reservists and new recruits. Craddock had left behind the *Canopus* because he believed she could not make more than twelve knots.

Craddock cruised north with his inferior squadron and on October 31 heard the wireless transmissions of a single German light cruiser *Leipzig* and deployed his ships in a search line and continued up the coast. He did not expect to encounter von Spee's combined fleet, which he imagined would be spread out along the coast, raiding commercial freighters as far north as the Galapagos Islands or the Panama Canal. In the previous few days, however, von Spee had received reports from German agents notifying him of British ships heading north along the southern coast of Chile, and he had kept his squadron together in hopes of overpowering and capturing British ships sailing alone. He wrote in his war journal: "The presence of strong enemy forces on the coast makes it impossible for the present for the Squadron to carry out its original intention of carrying on a war against commerce. This purpose is therefore renounced and the destruction of the enemy forces is substituted for it."

Ironically, even though von Spee was aware of Craddock's squadron, he only anticipated encountering a single ship, *Glasgow*, whose wireless transmissions near Coronel he had intercepted on November 1. He did not know the exact strength of the British squadron, nor did he expect all the British ships to be steaming together. Wireless radio technology was in its infancy, unpredictable and confusing. Radar did not yet exist, and intercepted snippets of wireless communication could rarely be trusted to be entirely accurate. Von Spee had craftily ordered radio silence on all the German cruisers except the *Leipzig* to deliberately give the impression that the ship steamed alone and not in cohort with the entire German squadron. Had Craddock suspected that the entire powerful German squadron was nearby, he would not have attacked. Certainly he would have waited for the *Canopus* to join him (although being so old, slow, and outgunned, what benefit she would have provided is debatable), and indeed he might have stayed clear altogether, waiting for reinforcements.

As the German squadron closed on the *Glasgow*, the British ship spied their smoke plumes on the horizon, picked up communications between the vessels, and hastily retreated to inform Craddock of the presence of several German ships. Aware of each other at last, the two squadrons spread out in a battle line in the late afternoon on November 1. Craddock attempted to close for battle quickly, to fight with the sun behind him blinding the German gunners, but von Spee, with speedier ships, kept out of range of the British guns and delayed the battle until the sun had set, thereby silhouetting the British ships against the last light of the day—illuminating them as targets while keeping the German ships in inscrutable darkness. Von Spee's squadron of five ships was vastly superior to the four British vessels. The German ships were modern, fast, heavily armed, and manned by crack crews of highly trained professionals as opposed to inexperienced reservists. They had double the broadside weight, longer-range guns, and much quicker firing rate.

The battle was decisive and quick. By darkness, it was all over. Just before eight P.M. the night sky was lit by a column of flame and smoke from a mighty explosion that shattered the British flagship *Good Hope*. British lieutenant Hirst observed in horror from the deck of the nearby light cruiser *Glasgow*, which was desperately fleeing the scene, that the *Good Hope* was reduced to "a low, black hull gutted of her upper works and only lighted by a dull red glare which shortly disappeared. Although no one on board *Glasgow* actually saw her founder, she could not have survived the shock more than a few minutes." Another British cruiser, *Monmouth*, was also aflame, and after additional pummeling by the heavy guns of the German cruisers, her broken metal hulk split, gushed full of water, heeled over, and was sucked beneath the cold water, her flag still flying. *Glasgow* was severely damaged and limped away into the darkness along with the armed merchant ship *Otranto*.

Von Spee's ships sustained little damage, while both British armored cruisers were sunk with all hands either killed by shell explosions or drowned in darkness in the frigid South Pacific. All told, nearly 1,600 British sailors were killed, including Craddock and all the officers of the two armored cruisers. The armed merchant ship *Otranto* and the *Glasgow*, after sustaining considerable damage, fled the battle to warn the approaching *Canopus* of the danger. These three surviving British ships fled back around South America to the Atlantic and the Falkland Islands to regroup—they could do nothing against such an overwhelmingly superior force. It was the Royal Navy's first significant defeat in over a century, leaving naval control over the west coast of South America, and hence the nitrate supply, temporarily in German hands.

Meanwhile, von Spee led his ships to the Chilean port of Valparaiso for reprovisioning. Although Chile was officially neutral in the global conflict, the rules of war provided that the ships of combatants could spend twenty-four hours in neutral ports to reprovision and load up on coal. Von Spee was hailed as a hero by the sizable German community and was presented with three hundred Iron Crosses for his officers. He also took on 127 eager seamen volunteers. He met with the German ambassador and the consul general, receiving updated reports on the war and the naval situation. He was warned of the converging enemy warships in the Pacific, West Indies, and South Atlantic and he was advised to "break through for home" while he still could. He probably knew he would be hunted by the Royal Navy—a victory such as he had achieved could not go unavenged. A Japanese-Anglo fleet was moving south from the Galapagos to prevent him from fleeing north; British and Australian ships were waiting for him in the South Pacific; and British and French ships were stationed in the Caribbean anticipating his use of the Panama Canal. Von Spee did not know exactly where they were stationed or how powerful they were, but common sense and intuition suggested that any squadron of ships sent against him now would not be weak enough to allow him a second easy victory.

Although von Spee knew he had to leave his South American haunt, he and the squadron spent most of the next month off the coast of Chile. His presence brought most commercial shipping to a halt. The historian Williams Haynes writes in *American Chemical Industry: The World War I Period* that von Spee "established to complete a control over the Chilean coast that British cargoes became uninsurable and nitrate shipments were practically suspended." Commercial shipping was confined to port for fear of attack or capture by the roving cruisers during the crucial fall season when large nitrate orders were traditionally shipped in advance of the spring planting season. In an era of incomplete news and erratic communication, it was impossible for merchant ships to know that von Spee was already far south of the main shipping routes and was planning a dash around the Horn into the Atlantic rather than launching an attack on commerce.

Von Spee was far from his home port, running low on ammunition, and his options were limited—there were no other German ships at large to offer aid. Realizing the precarious logistics of his situation, the German Admiralty had informed him that further cruiser warfare should be discontinued and that he should steam home while he could. Without access to bases to load up on coal, he had to travel in convoy with a retinue of seven heavily loaded and cumbersome colliers and rely on secret rendezvous with German-sympathetic local ships (the primitive engines of the steam-powered cruisers consumed vast quantities of coal such that a cruiser would normally have to replenish coal supplies every week at sea). In late November, leaving a small provision tender at the island of Mas Auera to give the false impression that his entire squadron remained in the vicinity, von Spee led his five warships and colliers south. They rounded Cape

Horn on December 6, headed into the Atlantic, where they would have several choices of route to Germany.

Three days after the battle news of the terrible defeat at Coronel reached London, cabled from Valparaiso, and it was soon confirmed by a wireless transmission from the fleeing light cruiser *Glasgow*. There was much hand-wringing at the Admiralty and a vociferous public outcry: How could the Royal Navy have suffered such a humiliating defeat? Who was responsible? How could the Admiralty have ordered an inferior squadron of outclassed and aging ships to attack a superior German force? And von Spee's squadron was still at large. The potential future damage that von Spee's roving squadron might cause was enormous; the potential further damage to the Royal Navy's reputation was worse. Whether the government was concerned about the destabilizing of the nitrate shipments is not recorded, but given the shortages of ammunition in the early months of the war, it ought to have been an issue. The loss of sixteen hundred men was significant, but to the war effort they were mostly untrained reservists, and the two aging cruisers were of little material value in the grand scheme of the conflict. But psychologically the defeat was a serious blow. Eliminating von Spee became a priority for the Royal Navy, and von Spee probably knew it. As he left Valparaiso on November 2, a woman stepped forward and presented him with a bouquet of flowers to celebrate his victory. "Thank you," he reputedly responded, "they will do nicely for my grave."

Within days of the battle, thirty-nine-year-old Winston Churchill, the political head of the navy as First Lord of the Admiralty, and the seventy-three-year-old bombastic ex-admiral Sir John Fisher, the professional head of the navy as First Sea Lord, had organized a relief and revenge expedition. The British perceived several possible courses for von Spee's deadly squadron and they had to be prepared for all of them: He might steam north to Panama to continue damaging merchant shipping; and

Admiral Graf von Spee shown in Valparaiso after the battle of Coronel in November 1914. Von Spee's defeat of the British fleet under Admiral Christopher Craddock temporarily placed naval control over the South American coast in German hands and disrupted the autumn nitrate shipments from Chile.

after passing through the canal, he might commence raiding in the Caribbean; he might turn west to disrupt shipping and troop transport in the South Pacific; he might head to South Africa and aid a German-friendly uprising by the Boers; or he might follow the fleeing British ships around the Horn of South America and raid the British colony, coaling depot, and wireless station at the Falkland Islands. Von Spee was a wild card that continued to take out of commission a great number of British and Allied ships as they had to defend against each of the numerous possibilities. Von Spee's five cruisers, and the two others he had detached from his squadron months earlier, effectively occupied five times as many Allied vessels, keeping them away from the critical zones of conflict in the North Sea and the Mediterranean. As the military historian John Keegan remarks in *Intelligence in War*: "The strategic return on the ratio was considerable." Eliminating these commerce raiders quickly would not only free up countless ships but, most importantly, would secure an undisputed control of the shipping routes to secure the transport of nitrates, agricultural products, and other needed raw materials. With nearly nine thousand steamships worldwide, Britain was by far the greatest maritime merchant nation on the globe, and London, as the center of global communications with its telegraphic cables, was the world's financial center. A substantial bulk of global trade involved either British firms or ships or was financed by British banking. As the nation with the greatest number of ships and the greatest involvement in global trade networks, Britain had the most to lose from commerce-destroying cruiser squadrons such as von Spee's. Von Spee threatened commerce from the western United States and Asia that passed through the Panama Canal, and most important, the indispensable Chilean nitrates, 60 percent of which were carried in British ships. If he reached the Atlantic he would also put at risk vital beef products destined for Britain from Argentina.

The British fleet that Churchill and Fisher assembled with the objective of hunting von Spee, avenging the battered pride of the Royal Navy, and reclaiming naval control over the South Atlantic and South Pacific, was indeed powerful. It consisted of six ships, the mighty battle cruisers *Invincible* and *Inflexible,* the armored cruisers *Cornwall, Kent,* and *Carnarvon,* and the light cruiser *Bristol.* It was decided that von Spee's most likely destination would be the Falkland Islands, and these modern cruisers, with well-trained troops under the command of Admiral Sir Doveton Sturdee, were to congregate into a squadron in South American waters near the River Plate, pick up the recently repaired *Glasgow,* and proceed down the east coast of South America and rendezvous with the *Canopus* at Port Stanley. Then they would hope for intelligence as to von Spee's location and begin the hunt. It was von Spee, however, who would find them once again.

Perhaps frustrated that logistics should force him to abandon his initial plan to prey on commerce before he had actually destroyed a single transport ship, or hoping for one last victory before returning home, von Spee, against the good judgment of several of his officers, planned an assault on Port Stanley at the Falkland Islands. After seizing control of the port, he planned to recoal before burning the remaining coal supplies, destroying the wireless station, and capturing the governor. What he planned to do afterward remains a mystery, since returning to Germany through British-controlled waterways would have proven difficult, and he had already received a wireless communication that the German fleet would be unable to offer him aid in the North Sea. Perhaps he intended to return to Chile or raid British commerce farther north around the River Plate in Argentina. Regardless of his future plans, his decision to attack the Falklands was based on faulty intelligence. The most up-to-date information available to von Spee, from erratic wireless communication and a meeting with a sympathetic ship encountered near Cape Horn, was that, as of December 6, the only British military ship in Port Stanley was the old *Canopus*, grounded defensively to guard the harbor entrance. Von Spee suspected that many British ships would be in Africa or around the River Plate. Certainly he did not anticipate that there could be a British fleet in Port Stanley, and on December 7 the four German cruisers headed northeast to a decisive naval battle that would not only seal their fate, but settle for the next several years the question of the control of the global nitrate supply.

Keegan writes in *Intelligence in War* that von Spee suffered a "failure of judgement; perhaps he had been too long at sea, too long in the loneliness of command." There was little to be gained by attacking the Falklands, and the ill-conceived scheme "was to result in the destruction of the East Asiatic Cruiser Squadron, in circumstances horribly equivalent to those of its victory over Admiral Craddock, his ships and men." By pure coincidence, the British squadron under Sir Doveton Sturdee, which had been maintaining radio silence as they cruised south, had arrived in Port Stanley on December 7—two days after the ship that von Spee encountered off Cape Horn had departed the harbor. On the morning of December 8, when the German cruisers drew near without precautions, they were spotted and chased and blasted apart by the more powerful British ships, with little damage and no loss of ships to the British squadron. Nearly two thousand German sailors were killed in combat or drowned, including von Spee and his two sons. Only the light cruiser *Dresden* escaped, fleeing southwest to the Argentinian and Chilean coast, eventually to be caught and sunk near Coronel several months later. Although the German command knew that von Spee's squadron, at large without a safe haven, would eventually have been captured or destroyed, the ignominious and fruitless defeat at the Battle of the Falklands was an anticlimactic ending to von Spee's initial triumphant victory.

With the destruction of the German cruisers at the Battle of the Falklands, the Allies enjoyed a considerable boost in morale; the prestige and reputation of the Royal Navy had been reclaimed; and the victory the German navy gained at Coronel was entirely nullified. Churchill observed afterward: "The strain was everywhere relaxed. All our enterprises, whether of war or commerce, proceeded without the slightest hindrance. Within twenty-four hours, orders were sent to a score of British ships to return to home waters. For the first time we saw ourselves possessed of immense surpluses of ships of certain classes, or trained men, and of naval supplies of all kinds." Although the strategic significance of the battle was not great—it only marginally shifted the balance between British and German naval power and had no effect at all upon the land campaigns in Europe—for the next several years, until the unrestricted U-boat campaign in early 1917, Allied dominance of South American, indeed global, trade was complete. Although the victory at the Battle of the Falklands was psychological rather than concrete, it enabled the Allied naval blockade of Germany.

The blockade of Germany was intended to serve a similar purpose to the naval blockade of Napoleonic France a century earlier—to starve German industry of vital raw materials, to deprive the German people and armies of foreign food products, and to isolate Germany from communication and commerce with her colonies and neutral nations. Trading-with-the-enemy legislation effectively stopped all direct shipments of raw materials and manufactured goods to Germany by making any vessel that was doing business with Germany or that was financed by German enterprise a legitimate military target. A Chilean nitrate company, for example, could have its transport ships searched at sea and the caliche nitrate seized if it was en route to Germany or German allies. Similarly, British firms were forbidden to provide services to Germans even in neutral countries, such as allowing German nitrate operations in Chile to transport their goods on British-controlled railways. The vast number of British ships controlled shipping routes, particularly, as during the Napoleonic Wars, the seaways to Europe through the English Channel, the North Sea, and the Mediterranean Sea. "On the one hand," writes the naval historian Thomas G. Frothingham, "the Entente Allies, by their conceded control of the waterways of the world, were, from the start, enabled to move men and supplies for their armies by means of these greatest common carriers of all the ages. By the same means they could supply and maintain their nations and their shipbuilding. More than this, they also had access to the neutral nations, to use all the advantages of agricultural and producing countries, which had cheap labor and were free from war conditions. On the other hand, the Central Powers were shut off from these advantages." The blockade effectively denied Chilean nitrates to Germany throughout the war. Without nitrates, upon which all the industrialized nations depended, agricultural harvests would decline and they would soon deplete their stockpile of munitions. Without munitions modern war would be impossible.

In the years before the war, Germany was by far the greatest importer of Chilean nitrates, consuming nearly 40 percent of Chile's annual production. In 1912, this amounted to an incredible 911,962 tons. German's nitrate imports, which supplied more than half of the nation's total nitrate requirements, were double those of the United States, three times the quantity imported by France, and seven times Britain's own imports. Germany was more dependent than any other nation on Chilean nitrates for industry and agriculture, and had made no particular attempt to stock up on nitrates before the war. Twenty thousand tons of nitrates were captured in Antwerp during an advance into Belgium early in the war, but even this amount was quickly consumed, raising the specter of nitrate shortages.

Neither Germany, nor England, nor France was truly prepared for the conflict with regard to nitrate supplies. The prevailing belief was that the war would be short, and no contingency plans were made for the supply of raw materials such as nitrates. Explosives were consumed with alarming speed by both sides in the early months of the war. The historian G. I. Brown has observed: "More shells had been fired in a thirty-five-minute barrage at the start of the battle of Neuve Chapelle than in the whole of the Boer War." This new style of modern warfare, with its emphasis on big mortars and bombs and mines, required far greater quantities of nitrates for explosives than in any previous conflict. The British nitrate shortage came to a head in early 1915 due to ill planning and the logistical problems associated with shipping vital raw material from the far side of the world.

In the early years of the war, British troops were chronically

The British battle cruiser Inflexible *retrieved survivors from the German cruiser* Gneisenau *after the Battle of the Falkland Islands in December 1914.*

short of artillery and high explosives, and the shortage contributed greatly to the deaths of thousands of soldiers and rendered ineffective military plans and operations. G. I. Brown writes that at the start of the war "the total amount of high explosive available would have been laughable if the situation were not so serious. While the Germans were actually firing off about 2,500 tonnes of TNT every week, the total production of both TNT and Lyddite in Britain was less than 20 tonnes per week." Sir John French, the commander in chief of the British forces in Europe, repeatedly requested better and stronger explosives for his armies. He wrote in frustration after the Battle of Aubers Ridge in May 1915: "After all our demands, less than 8 per cent of our shells were high explosive, and we had only sufficient supply for about 40 minutes artillery preparation for this attack. . . . As I watched the Aubers Ridge, I clearly saw the great inequality of the artillery duels, and, as attack after attack failed, I could see the absence of sufficient artillery support was doubling and trebling our losses in men." David Lloyd George in his *War Memoirs* included several letters from soldiers on the battlefield. "We have just to sit tight in the trenches," wrote one, "while the German high explosives shatter them to bits." Another wrote: "It makes your heart break to see those men going forward and then held up—one after another, fighting and struggling, until, wearied out, they collapse like a wet cloth. And why? Because there is not an adequate supply of high explosives to blow the wire to bits and let our men get at the enemy." The shell-shortage scandal caused the fall of Herbert Asquith's Liberal government on May 26, 1915, which was replaced by a coalition ten days later and prompted the formation of the new Ministry of Munitions, headed by Lloyd George and later by Winston Churchill, tasked with the job of increasing the production of high explosives, which necessitated importing greater quantities of nitrates from Chile. During the war, the increased demands for Chilean nitrates from Britain, France, Italy, and later the United States, more than made up for the loss of the German market, rising steadily from 1,847,000 tons in 1914 to a record 2,919,000 tons in 1918.

Despite British control of the seas and the blockade of the German coast, Allied nitrate supplies were never entirely secure. At the time, the nitrate fleet consisted of an incredible 105 steamers and 23 sailing vessels, double the number of ships before the war, such were the quantities of nitrates routinely being shipped overseas. In early 1917, the sinking of several freighters by U-boats in the English Channel and off the coast of France caused a severe shortage of French nitrate supplies in advance of an anticipated German offensive in the spring. The French requested an emergency shipment of at least 75,000 tons, which Britain could only partially meet. About 12,000 tons of nitrates from the U.S. Navy was loaded onto a transport ship, but the ship was damaged in convoy and had to return to port, while, as Williams Haynes noted, "Another burnt in Halifax harbor under suspicious circumstances." The Nitrate Section of the U.S. War Industries Board hastily commandeered a steamship in the Panama Canal, along with 42,000 tons of Chilean nitrates that ostensibly belonged to the Department of Agriculture, and rushed it across the Atlantic just in time to prevent a shutdown of French explosives factories. At the same time, political interference in the form of labor disruptions in the nitrate regions in Chile and at the ports of Valparaiso, Antofagasta, and Iquique organized by German-financed agitators temporarily disrupted the supply.

The entry of the United States into the war in April 1917 on the side of Britain, France, and Italy placed incredible demands on the nitrate supply and the nitrate fleet. All the thousands of troops being sent across the Atlantic needed ammunition, while the needs for agriculture also increased to produce more food for the war effort. For several months before it was obvious, or a good gamble, that the United States would enter the war, thereby increasing American nitrate requirements once war was declared, speculators began hedging by purchasing stockpiles of nitrates in anticipation of the increased demand and rising prices. The price for nitrates did spike: By April the spot price for nitrates had risen to 7.5 cents per pound, an astonishing 300 percent increase, driven by the flurry of purchases on speculation and by frantic French and British purchasing agents fearing a shortage. Bernard Baruch, head of the Raw Materials Division of the War Industries Board, and his technical advisor Leland Summers, recognizing the vital necessity of a regular supply of nitrates at predictable prices, devised a bold scheme to keep a lid on escalating prices and to prevent speculators from disrupting the stability of the market.

They let it be known that for all U.S. government contracts, armament manufacturers could depend on a price of 4.5 cents per pound. With the price of nitrates guaranteed by the government, munitions manufacturers relaxed and ceased bidding and the price subsequently stabilized. The speculators who hoped for a quick profit, and the Chilean producers who were more than happy with the higher prices, were justifiably worried and wondered how the Raw Materials Division planned to secure the nitrates at the lower price. Had a new source of nitrates been discovered? Or had chemists devised an artificial means of manufacturing nitrates? If so, they could stand to lose a fortune. Baruch and Summers actually had no idea where they would obtain the cheap nitrates to fulfil the price guarantee, but good fortune intervened. The Chilean government had on deposit in Berlin approximately $17,000,000 in gold bullion, part of the nation's reserve guaranteeing its paper currency. Chile was in need of the additional gold to back increased foreign purchases and requested its return. Germany promptly refused. U.S. Navy intelligence officers picked up portions of the discussions and then, through diplomatic channels, let it be known that the German refusal to return the Chilean gold was a legitimate excuse to confiscate the excess supplies of nitrates from German-owned businesses within Chile. German nitrate refineries had amassed 235,000 tons of nitrates that could not be shipped, even for sale to Allied or neutral nations, because of the British Trading-with-the-Enemy Act, which denied them access to jute bags from India and prohibited transportation on British-controlled railroads. The United States offered to pay the Chilean government in gold bullion for the confiscated German nitrates.

Winston Churchill, newly shifted from the Admiralty to the Ministry of Munitions in the British government, quickly arranged for jute bags and for transportation on British railroads. Within a few months, the confiscated nitrates had been delivered to the United States and the gold bullion had arrived in Chile. Churchill also agreed to collude with Baruch by publicly declaring that Britain was out of the market for nitrates at the same time that Baruch and the War Industries Board made good on the offer of below-market-value nitrates for use in government contracts. They were selling the confiscated German nitrates at a little more than they had paid the Chilean government for them. This caused panic among the speculators, who began dumping their stocks, believing there was somehow a great excess of nitrates, and the price plummeted back within the usual range. Out of the spirit of cooperation and goodwill between Baruch and Churchill, and their success in manipulating nitrate prices, came the realization that past ad-hoc purchasing practices were much too open to disruption and interference and were not suitable for meeting wartime demands. Baruch wrote in *American Industry in the War: A Report of the War Industries Board* that "it was clearly necessary to eliminate this haphazard competitive system of buying." The Chilean monopoly on production had to be counterbalanced by an Allied monopoly on purchasing to prevent the upward creep of prices when they competed and bid against one another. By December 1917, France, Italy, Britain, and the United States formed the International Nitrate Executive, with its offices in London. Its responsibility was to purchase Chilean nitrates for the Allies and distribute the stock among each country according to need. All the prices paid were pooled and averaged in an attempt to provide some predictability to the volatility of the nitrate market and stabilize the supply of this vital raw material.

Although the International Nitrate Executive managed to smooth out many of the wrinkles in the supply of Chilean nitrates for the war, the problem could never be truly solved so long as the only source of the raw material was in a distant foreign country. Anxiety persisted throughout the remaining months of the war—transport ships could be sunk, the refineries or railways in Chile could be sabotaged, or the Chilean government could withhold shipment (although this last was unlikely, given that nitrate export revenue floated the government, and it would have effectively amounted to a declaration of war).

The Chilean caliche nitrate industry was predominantly foreign owned, owing to the high capital investment necessary to extract and purify the caliche on a profitable industrial scale and the need for imported foreign goods such as coal. British firms were at the center of this lucrative two-way trade. They imported coal and other manufactured products into Chile, accounting for 40 percent of Chile's imports, while shipping nearly 60 percent of Chile's nitrates to Europe and the United States. The quantities of caliche nitrate exported had increased steadily throughout the nineteenth and early twentieth centuries. Between 1896, the year of Alfred Nobel's death, and 1913, total nitrate exports had more than doubled, to an incredible 2,738,000 tons. Chile's caliche nitrate exports supplied over two-thirds of the global nitrate demands, while the tax revenue it generated constituted more than 60 percent of the nation's entire budget. This seemingly endless geyser of money was lavishly strewn about, financing grand public works, railroads, monuments, roads, ports, and a burgeoning public service. Chile was as dependent upon the foreign income from the nitrate trade as Britain, France, and the United States were dependent upon the nitrates for their war effort.

But in June 1918, a sympathetic Chilean court supported a legal action brought by the German nitrate companies against British import firms to recover fuel oil that had been contracted for but not delivered because of the Trading-with-the-Enemy Act. The court granted an injunction that prohibited the British traders from supplying oil, which had recently replaced coal as a fuel source, to anyone else before fulfilling the contracts with the German nitrate firms. Since the nitrate refineries were powered by oil, the sudden loss of this energy source could have paralyzed many operations for months. Fortunately for the Allies, most of the oil in Chile actually originated in California, and when the U.S. oil companies, under the guidance of the War Industries Board, threatened to withhold all oil shipments to Chile unless the injunction was eliminated, the Chilean government became involved and set it aside.

The Allies were constantly worried about political interference of this nature. Although no similar incidents arose for the remainder of the war, no one could know it at the time. It was a game of brinksmanship and continuous jockeying for the upper hand as the Chilean government's twin desires for maximum profit and a tepid sympathy for German interests competed with their desperate need for a stable flow of foreign currency. There were many German sympathizers in Chile. The only way the United States, Britain, and France would ever be liberated from the Chilean monopoly on nitrate would be to devise a method for manufacturing synthetic nitrogen, and great efforts were made in the final years of the war to achieve nitrogen self-sufficiency through various experimental methods. The need for nitrates for modern explosives and the precarious access to the only commercially viable natural supply was the Achilles' heel of the Allied war plans, a problem that was never completely solved. "All the men and all the cannon America might bring to Armageddon," wrote Grosvenor Clarkson just after the war, "would be powerless if the rusty tramp steamers could not maintain their drab procession from Chile to the ports of America and the Allies." The U.S. War Industries Board remained apprehensive about the nitrate situation, "forever haunted by the specter of a war won or lost by some mishap in the far-away Chilean deserts or in the waters between."

"Throughout the war," writes Williams Haynes, "American farmers called upon to help feed Europe were chronically short of this vital plant food, and at critical junctures serious shortages of ammunition, due to lack of nitrogen, created hazardous positions at the battle front. The need for nitrogen sat like Banquo's ghost at every Allied council table. It forced many an anxious decision."

Not so for Germany. Some historians have argued that Germany did not enter the war until it had found a secure and unlimited source of nitrogen. While this might be giving too much credit to the military bureaucracy, and it is an assertion disputed by others, the timing of one of Germany's greatest scientific breakthroughs came suspiciously close on the heels of the war.

The Father of the War

Fritz Haber's World-Changing Discovery

A man belongs to the world in times of peace, but to his country in times of war.
                                        —Fritz Haber, ca. 1916

At the time of his greatest scientific triumph, Fritz Haber was a short, plump, balding man with prominent round spectacles and a neatly trimmed mustache. His face showed well-worn creases, giving him a stern expression. He was frequently photographed or drawn with a large lit cigar in hand, which he reputedly left burning on his laboratory bench while engrossed in his work. He was a narrowly focused, ambitious, and competitive scientist. His portraits and photographs, however, never hint at his livelier charismatic antics as a storyteller and amateur dramatist. During the First World War he worked with an officer who habitually strutted about the office in riding spurs. Amused at the man's posturing, one day Haber asked him to "jump on your horse and ride into the next room for the documents." The enigmatic Haber was a gregarious and highly social man who took a genuine pleasure in public events, lectures, and performances. He was a well-rounded and broadly educated polylinguist, versed in classical literature and art. Yet ambitious competitiveness and deference to authority were his strongest traits, and he was driven to succeed at his scientific pursuits, never settling for second best.

He worked long hours in his laboratory and spent little time at home. Periods of animated and forceful discourse on scientific questions were juxtaposed with quiet rumination and introspection. Toward his family, he could be overbearing and autocratic, leading to a deeply unhappy marriage that ended in tragedy. He bullied his eldest son into studying chemistry instead of law. But while he was a poor husband and father, in the scientific community his colleagues remembered him fondly and with respect. He was an admired, perceptive, and empathetic mentor, and students from around the world came to study under his guidance before and after the war. A sketch of him in his army uniform during the First World War hints at the fierce duty, loyalty, and patriotism that characterized his wartime activities—activities that earned him the accolades of a national hero in Germany but near universal condemnation abroad. In 1919 he was awarded a Nobel Prize in Chemistry for a brilliant scientific discovery that, like many other discoveries and developments in the curious history of explosives, was a powerful tool for the seemingly unrelated activities of construction and destruction, of creating and ending life. Controversy attended his award, just as controversy has dogged his reputation long after his death as a lonely exile in 1934.

Born in Breslau in Prussia in 1868, the first son of a prosperous Jewish chemical wholesale merchant dealing in natural dyes, paints, lacquers, and other chemicals and drugs, Haber traveled an unlikely road to become one of the most prominent and distinguished German chemists in an era known for great chemists and chemical breakthroughs. His mother died in childbirth, and he was raised by relatives for several years until his father remarried; eventually he had three half sisters. Throughout his childhood he was given every opportunity to pursue his interests and he showed great affinity for languages, theater, and literature. He traveled throughout Europe on numerous family vacations and was encouraged to study and learn whenever possible. When he first enrolled at the University of Berlin in 1886 his course of study was broad, but he soon took a special interest in chemistry. Without financial worries, at the end of his first semester he left for Heidelberg University, where he continued studying chemistry and physics. Under the tutelage of Professor Robert Bunsen he was encouraged to seek a practical rather than theoretical focus to his studies, advice and training that had an impact on the rest of his career. Haber always sought practical applications for his work or set out to solve problems that would have an impact on the world in a tangible way. The young Haber returned to Breslau one year later for compulsory military service in the artillery, and then enrolled at the Charlottenburg technical institute in Berlin, from which he earned a degree of Doctor of Philosophy in 1891 when he was twenty-three years old. After graduating he worked and studied at several business and educational institutes around Europe before returning to Breslau to join his father's firm. But he found the life of a traveling salesman to be dreary and unchallenging, and he and his father frequently clashed over his role and responsibilities in the business. After less than a year he heeded his father's advice to "Go to a university. You don't belong in business."

Haber traveled to the University of Jena where he spent an unsatisfying year and a half earning some money and gaining experience, before settling at the Karlsruhe technical institute as an assistant in the Department of Chemical and Fuel Technology. At Karlsruhe he thrived, and after several years of hard work, during which he published several well-received books and numerous articles, he was promoted to professor of physical chemistry and director of the Electrochemistry Institute. It was here he would accomplish the greatest scientific breakthrough of his distinguished career. In 1902 he was sent on a tour of the United States to visit universities and chemical plants and report on the state of the American chemical industry. In his typically blunt and critical manner he concluded that the American chemical industry and teaching infrastructure was primitive compared to the German industry, which was the most advanced in the world at the time. On one of his visits to American chemical operations he toured, prophetically, the Atmospheric Products Company at Niagara Falls, which was using the electric arc process for ammonia synthesis—from which nitric acid can be made. Although it was not his specialty, Haber observed technical problems with the operation and wrote a critical report outlining some of these difficulties.

In 1904 Haber was approached by a Vienna chemical company to do his own research on ammonia synthesis, but after considerable theoretical study and laboratory testing on the problem, and an analysis of the current state of scientific knowledge, he wrote them with inconclusive results. Meanwhile he continued to investigate a wide range of topics, including developing electrodes for calculating the acidity of liquids, studying the energy loss in engines, and analyzing the properties of the Bunsen flame. He was a prolific writer and published over fifty scientific papers between 1900 and 1905. Several years later, in 1907, Haber returned to the problem of ammonia synthesis, with his characteristic intensity and focus. Typically for Haber, he only turned his full attention to the problem when he felt slighted by another distinguished scientist, Walther Nernst, at the University of Berlin, who challenged some of his early conclusions and described his work as incorrect. Nernst had just published his theory of thermodynamics, for which he later received a Nobel Prize in Physics in 1920, and he observed that Haber's data from his ammonia synthesis experiments years earlier contradicted his own theoretical predictions. Haber then repeated his ammonia synthesis experiment and came up with slightly different data, but these again did not correspond with Nernst's predictions. Nernst then publicly criticized Haber at the annual meeting of the German Bunsen Society in May 1907.

Haber was devastated. For a person as distinguished in the scientific community as Nernst to publicly denigrate his work was seriously damaging to his reputation as a new professor. The stress

This is a portrait of Fritz Haber, the enigmatic German chemist whose discovery of a method for creating synthetic nitrogen kept Germany fighting in the First World War. Since then his discovery has subtly yet radically altered the world by supplying unlimited nitrates for explosives and fertilizers.

caused his health to deteriorate, but instead of crushing his spirit it inspired him to focus his mind on the problem of ammonia synthesis to prove Nernst wrong. In his laboratory Haber and his assistant built an apparatus for continued experimentation—an iron canister 75 centimeters (30 inches) tall that could withstand the extreme heat and pressure needed for their experiments. Creating the apparatus required overcoming several technical hurdles, the most difficult being the creation of valves that could regulate the flow of gasses under the extreme conditions needed for the reaction to succeed.

The two gasses, hydrogen and nitrogen, were pumped through the entrance valves into the pressurized canister and were superheated by a nickel heating coil and were circulated by a small primitive pump. The catalyst to the ammonia synthesis reaction was held in position near the gas exit valve, and as the gasses passed out the exit the catalyst completed the reaction. After leaving the heated and pressurized canister the gasses cooled rapidly and formed powdered ammonia. After designing his small workable experimental converter, Haber and his assistants spent months repeating the procedure with different balances of temperature and pressure, different quantities of nitrogen and hydrogen, and different catalysts. To minimize the energy needed to complete the reaction, they sought to reduce both the temperature and pressure needed for the reaction so as to achieve an economically viable process. They tested numerous catalysts, including calcium, manganese, nickel, osmium, before settling on uranium as the most effective substance. It best served the dual purpose of increasing the yield of ammonia and reducing the temperature required for synthesis. Haber was not interested in merely showing that the reaction could be done, thereby creating a technical and scientific curiosity and a theoretical defense of scientific principles. He was much more ambitious. From the outset he sought to establish a process that would have the potential to become an industry. Haber's objective was to solve the nitrogen problem for agriculture and explosives, and by mid-1909 he was ready to demonstrate that he had done so.

In July 1909, Haber and his assistants had optimized the pressure and temperature ratio, settled on a catalyst, and constructed a model apparatus that promised to create synthetic ammonia using an entirely new procedure that was far more efficient than anything previously conceived. He gave a demonstration in his laboratory to two members of the giant chemical firm Badische Anilin und Soda Fabrik (BASF), the research engineer Carl Bosch and the chemist Alwin Mittasch. After a technical glitch that delayed the demonstration for several hours, Haber's apparatus produced for his skeptical visitors a small quantity of powdered synthetic ammonia. Haber's patent description outlines in broad terms the nature of his process: "Process for synthetic production of ammonia from its elements, whereby an appropriate mixture of nitrogen and hydrogen is continuously subjected to both the production of ammonia under the influence of heated catalysts, and continuous removal of the resulting ammonia, characterized by constant pressure and the transfer of process heat from the ammonia-containing reaction gases to the incoming ammonia-free gas mixture." It was a discovery that would have profound implications—if the laboratory results could be expanded into a large-scale industrial process.

In the early twentieth century, the discovery of a method for artificially producing nitrogen compounds was one of the great technical puzzles of the age. At his presidential address to the British Association for the Advancement of Science in 1893, Sir William Crookes raised the specter of global starvation because of a lack of nitrogen for fertilizers. "The fixation of nitrogen," he claimed, "is vital to the progress of civilized humanity." He, inaccurately as it turned out, predicted that the supply of Chilean nitrates would be exhausted within twenty to thirty years. This shocking news was the impetus that inspired research into the problem. Nitrogen is common in the atmosphere: Approximately 80 percent of each cubic foot of air consists of nitrogen, and above every square yard of the earth there are seven tons of the gaseous element. So when scientists began searching for methods to reduce the dependence on Chilean nitrates, extracting nitrogen from the air was the obvious starting point. Nitrogen fixation—a method for extracting nitrogen from the air and converting it into nitrogen compounds—had produced more than three thousand scientific papers by 1915.

At the end of the nineteenth century, there were three experimental and imperfect methods for obtaining nitrogen compounds from sources other than natural deposits: the high-temperature cyanamide process, the electric arc process, and collection of ammonia by-products from the coking of coal. The cyanamide process involved combining nitrogen with calcium carbide at high temperatures. After several days of reaction in a contained environment, a black powder called cyanamide was produced. Sometimes it could be used directly as a fertilizer, but usually it required more processing. It was difficult to apply to fields because it was too powdery and it was a skin irritant. Although several plants were constructed to create cyanamide near sources of great hydroelectric power—in Piano d'Orta in Italy, near Cologne and Trosberg in Germany, and later in several other European countries, Japan, and at Niagara Falls in Canada—the high energy requirements, including high quality coke and great quantities of electricity, severely hindered its commercial application and uses. The process never supplied great quantities of nitrogen, even in Germany when war demands temporarily justified the high cost of production. The electric arc process, in which nitrogen and oxygen form nitric oxide in an electric arc, which is then converted into nitric acid for explosives, suffered from similar problems of economic viability. This was the process used by the Atmospheric Products Company at Niagara Falls that Haber visited in 1902 (and which went out of business in 1904). Although several years after his visit Haber improved upon the process, it still remained uneconomical. It was too energy dependent, and hence far too expensive. Several plants operated in Norway because of the abundant hydroelectric power, but the process never became truly economical on a large scale. It never produced even a fraction of the nitrogen created by the cyanamide process.

Collecting the nitrogen by-product from coking coal was perhaps the most useful in supplying great quantities of nitrogen in every country in the early twentieth century. Coal usually contains small quantities of nitrogen, not more than one to two per cent, that were "derived," according to Vaclav Smil in his comprehensive study of nitrogen and fertilizers *Enriching the Earth*, "from degradation of proteins that were present in the biomass whose eventual transformation by pressure and heat produced the solid fuel." When coal is burned, the nitrogen is released into the atmosphere "but when coal is heated in the absence of air—during production of coke needed for pig iron smelting, or during generation of coal gas, which was so widely used for lighting in nineteenth-century cities—part of the fuel nitrogen, typically 12–17 percent, is released as ammonia." After the replacement of the older, inefficient beehive ovens with by-product recovery coke ovens in the late nineteenth century, the raw coke oven gasses could be commercially captured, processed, and converted into ammonium sulfate, which in turn could be converted into nitric acid by the Ostwald process. The Ostwald process, perfected in 1902 by the German chemist Wilhelm Ostwald, who was awarded a Nobel Prize in Chemistry in 1909 for his discovery, converts ammonia into nitric acid using a catalyst at high temperature.

In the early twentieth century, each of the three commercial methods for obtaining nitrogen was unsuitable in its own way. One of Haber's biographers, Morris Goran, writes that "all the means devised to fix or combine nitrogen out of the air were faulty. Either the installation cost was exorbitant, the upkeep and repair of installations expensive, the process inefficient, raw materials not abundant, or the procedure only suitable where water power was plentiful." Even the most efficient method, recovering the ammonia by-product from coking, was limited by the amount of coal being transformed into coke. A great deal of research and experimentation went into increasing the quantities of ammonia recovered through the coking process, but no significant improvements were ever made. The quantity of ammonium sulfate generated was entirely dependent upon the amount of coal consumed in by-product recovery ovens. Although it was a significant quantity, it was static; output could never increase along with increases in demand. The conversion to by-product recovery ovens was much slower in the United States, and industry was still using antiquated and inefficient beehive ovens throughout most of the First

World War, severely reducing the quantity of ammonia produced. Vaclav Smil has noted that although coking by-product ammonium "eventually contributed relatively large amounts of nitrogen, none of them [the cyanamide or arc process] solved the problem of long-term provision of fixed nitrogen: they represented welcome additions of fixed nitrogen but not lasting solutions."

Haber's process for ammonia synthesis was entirely different. Although it was initially met with considerable criticism, Haber's discovery appeared to solve the problems associated with the cyanamide and arc process, particularly the need for vast and unreasonable quantities of energy. Although it posed other problems that would need to be solved before becoming commercially viable, notably creating huge iron vessels capable of withstanding the high pressures and temperatures required for obtaining the greatest yield, it was still a breakthrough. The historian G. I. Brown describes the basic and fundamentally simple principle behind Haber's process: "Nitrogen is obtained from the air," he writes, "and hydrogen from the water or methane, and the two gases are made to react at a temperature of 550 degrees centigrade, under a pressure of between 150 and 300 atmospheres, and in the presence of a catalyst, to form ammonia gas. . . . The process was something of a landmark, first because the conditions required to obtain a good yield of ammonia were worked out from first principles, and secondly because it was a breakthrough in high pressure technology. It was such a breakthrough, in fact, that it was at first regarded with much incredulity and suspicion by many industrial chemists." The reaction between nitrogen and hydrogen to form ammonia was not a new discovery. It had been common knowledge amongst chemists for years, but Haber's innovation was to drastically improve the yields and speed up the reaction time by creating the proper conditions of high heat, high pressure, and a suitable catalyst. Haber applied for a patent and quickly entered into an agreement with the huge chemical firm BASF for the exploitation of his remarkable technological innovation. But in 1909 Haber's model was only a small experimental converter in his lab at Karlsruhe. It would take a lot of work and an incredible financial investment to transform his model and to scale it up into a working pilot plant—to create an effective and economical commercial design to exploit the patent. Within BASF there was concern that the costs and time to work through these problems would be prohibitive, or perhaps even impossible. The monumental engineering challenge was assigned to Carl Bosch, a thirty-four-year-old metallurgical engineer with knowledge of the alternate methods of nitrogen fixation and the capabilities of the steel industry in working with high-pressure converters. Bosch undertook the daunting task of enlarging and perfecting Haber's design on an industrial scale with characteristic energy and creativity, and within a few years, by 1913, the first BASF chemical ammonia synthesis plant was running at Oppau. It could produce approximately 36,000 tons of ammonia, which converted into 7,200 tons of nitrogen, which was being used for fertilizers. "Then the war broke out," writes Vaclav Smil, "and the role of ammonia synthesis changed in just a matter of months from producing fertilizers to sustaining Germany's munitions industry."

Whether the old saying that necessity is the mother of invention is universally true, it was certainly the case in Germany during the early days of the First World War. Britain's naval blockade of Germany and German allied countries was slowly strangling them of raw materials from other countries, in particular Chilean nitrates, and the fear of being shut off from access to nitrates was an inspiration to the rapid industrialization of Haber's experimental design. Bosch had taken three years to overcome all the formidable technical problems and transform Haber's experimental laboratory apparatus into a large-scale, commercially viable plant at Oppau.

He was then tasked with rapidly expanding the existing enterprise to meet the demands of the German military, doubling or tripling the output as quickly as possible. Remarkably, he was successful, and the production of nitrogen doubled in 1915 and then quadrupled in 1916. By 1917 he had finished overseeing the construction of a second, even larger, plant at Leuna in central Germany.

Some historians have speculated that the German military had prior knowledge of Bosch's work at BASF and timed the declaration of war to coincide with the technical breakthrough. "So while Wilhelm II rattled his sword noisily for several years," wrote the American historian Williams Haynes in 1945 in his colorful style, "he did not throw away the scabbard until the Haber synthetic-ammonia process and its companion, the Ostwald process for the conversion of ammonia to nitric acid, had been perfected and tested." Likewise Professor of Chemistry J. E. Zanetti wrote in *The Significance of Nitrogen* in 1932: "It has been suggested that not until the Haber process had reached a commercial scale of operations did Germany dare to go to war. For with the practical certainty of being cut off from Chilean nitrate deposits by the British fleet, it would have been suicidal on her part to risk finding herself unable either to manufacture munitions or fertilize her fields." Most historians of the chemical industry and the war disagree, however, particularly more recent writers, pointing out that at the outset of the war Germany was as grossly unprepared for meeting the increased nitrogen demands as France or Britain. The German government appeared to have no detailed or intimate knowledge of the various methods of producing synthetic nitrates, and Bosch himself reported after a meeting with members of the War Ministry in September 1914 that they had a complete lack of appreciation for the chemical requirements of modern war. They had based their estimates of the quantity of nitrates needed for the conflict upon the quantities used during the Franco-Prussian War over four decades earlier. Fritz Haber's son, the economic historian L. F. Haber, writes in his comprehensive book *The Chemical Industry, 1900–1930:* "Given the scale of these requirements [in Germany], the nitrogen problem as it developed after August 1914 was not simply a choice between guns and grain, but between guns or defeat. Looking back, it is remarkable that despite all the warnings of an impending shortage, no practical steps had been taken to safeguard nitrogen supplies for military purposes."

The sudden shortage of nitrogen was certainly an impetus for action once the war began and the German high command was apprised of the dire shortages of raw materials that would result if a lightning victory was not achieved. The historian Jeffrey Allan Johnson writes in *The Kaiser's Chemists* that in a meeting in September 1914 the prominent German scientist Emil Fischer pointed out to the War Ministry the "critical gap between production and requirements for toluol (for TNT), ammonia, nitric acid, gasoline and oil, coal and coke . . . synthetic sources [of nitrates] covered less than a tenth of German needs in 1913." Johnson continues: "All the capabilities would have to be expanded as rapidly as possible if the German military machine were not to grind to a halt within a few months for lack of explosives." German cyanamide plants, still the preferred method for manufacturing nitrogen, were immediately expanded, and the industry had great hopes of finally becoming profitable with a monopoly position during the war. But it was the additional ammonia that unexpectedly came on line from the BASF plants at Oppau, and later at Leuna, using the Haber-Bosch process that kept German industry and German farmers supplied with the indispensable raw material of nitrogen.

Without Haber and Bosch, Germany would have run out of munitions, probably by the spring of 1916. Germany was counting on negotiating for peace after a quick land-based military victory and was unprepared, as were Britain and France, for a prolonged war of attrition. "In this manner," writes L. F. Haber, referring to the rapid expansion of production at Oppau and Leuna, "the development of the Haber-Bosch process between 1915 and 1918 enabled Germany to continue the war at home and at the front. . . . In 1917 and 1918 ammonia made by the Haber-Bosch process represented 45 and 50 percent respectively of the country's production of nitrogen compounds. Without this contribution, the war would undoubtedly have ended sooner." Nitrates from the Haber-Bosch process increased from a negligible quantity before the war to over two hundred thousand tons annually by the end of the war, supplying more nitrogen than the cyanamide process (in excess of 60 percent more) and exceeding the total nitrogen produced in all of Germany's coking plants for by-product recovery. Although many general war histories do not discuss the issue of nitrate supply as something of strategic importance during the conflict, historians of science and specifically the chemical industry have frequently discussed and studied the situation. As recently as 2001, in his dense and comprehensive study of fertilizers and the work of Haber and Bosch, Vaclav Smil concludes that "Oppau and Leuna made the decisive difference to Germany's capacity to produce munitions: their success had undoubtedly delayed the Second Reich's collapse."

While Germany had effectively solved the nitrogen problem during the war, the Entente Allies remained dependent upon shipments from Chile, by-product recovery from coking, and some piddling quantities derived from the cyanamide process. Throughout the war, scientists in Britain, France, Italy, and the United States attempted to replicate the results of the Haber-Bosch process. Several synthetic ammonia plants were actually constructed in the United States; the most significant were Nitrate Plant No.1 in Sheffield, Alabama, and Nitrate Plant No.2 in Muscle Shoals, Alabama, but neither produced any significant quantities during the war. A British plant at Billingham and a French plant at Toulouse likewise failed to produce any nitrogen. Although the Haber-Bosch process was simple in concept, it was technically difficult to create the proper high temperature and pressure environment for the reaction to occur. These countries also lacked knowledge of the catalysts needed to carry out the reaction. It was not until after the war that the secrets were revealed.

Fritz Haber was awarded a Nobel Prize in Chemistry for his remarkable discovery in 1919, the first postwar award in chemistry. During the years of 1916 and 1917, the awards were canceled. Twenty years later, Carl Bosch was also awarded a Nobel Prize for his role in adapting and expanding Haber's original model. Rather than the usual celebration and universal congratulations following Haber's award, however, it was met with outrage from scientists in Allied countries. At the time, it was an exceedingly contentious decision by the Nobel Committee. Haber's award is considered one of the most controversial of all the Nobel Prizes awarded in over a century, and was not presented to him in person by the king of Sweden, nor at the same time as the other awards (he received his award in the summer of 1920, without the usual fanfare). So great was the ill feeling that scientists from Allied countries refused to nominate a German for the chemistry award for the following twelve years. This aversion to Haber and disgust with his Nobel Prize arose not because they felt Haber was undeserving of recognition—indeed, his discovery was scientifically creative and of immense practical benefit to the world—but because of his other, more sinister, wartime activities.

In 1911, two years after demonstrating his ammonia synthesis model to BASF, Haber had left the Karlsruhe institute to head up the newly formed Kaiser Wilhelm Institute in Physical Chemistry and Electrochemistry in Berlin. Within months of the outbreak of the war, however, work was severely disrupted when many of the young employees were called into military service. Haber himself eagerly volunteered his services but was refused, and he drifted into depression after being denied the opportunity to serve his country. Within weeks, though, Haber's spirits were raised when the institute was enlisted to aid the war effort, and he became a soldier-scientist in the German military. Not surprisingly, Haber was consulted about nitrates, then under the care of Bosch at Oppau, and he was also set working to devise a gasoline with a low freezing point to be used during the Russian winter. But the most significant task that Haber, along with other prominent German scientists, was given in the final months of 1914 was to develop an effective method for using chemical agents to clear the trenches and force troops into the open where more traditional styles of warfare would prevail. It was not an original idea. British chemists had been experimenting with chemical agents for use in the war, and French troops had used small quantities of tear gas, but ineffectively. Early in the war, the Russians had dabbled with chlorine gas, but the effort was unsuccessful because the gas froze in the winter and sank into the snow, only to be released in the spring during the thaw, when enemy soldiers were long gone. Haber immediately recognized the potential of chlorine gas.

After several months of testing, and a devastating explosion in December 1914 that killed one of the scientists working under Haber and severely wounded another, he had settled on a means to release chlorine gas from cylinders. By the spring of 1915 Haber had organized the first gas assault, a series of well-positioned canisters along a three and a half mile front near Ypres. There was much debate as to whether using the gas was immoral or a violation of international law, and whether or not English and French forces would also develop the technology and use it against them. The prevailing winds on the western front did not favor Germany, and one of Haber's colleagues, Emil Fischer, hoped for "failure from the bottom of my patriotic heart; for if he succeeds, the French will soon figure it out and then turn the tables, which will be very easy for them to do." Haber, however, argued that there was nothing unchivalrous about gas weapons and that if he was successful, it would save lives by bringing the war to an end sooner. Haber's was an argument similar to the arguments put forward by Alfred Nobel to justify his life's work in improving the destructive capacity of explosives and propellant powders—that each new terrible manifestation of destructive power is a humane development that will shorten conflicts and therefore ultimately result in saving lives. As Nobel realized near the end of his life, it is a somewhat circular and self-deceptive argument that has never proven accurate. But at the same time, there was a sanctimonious and morally inconsistent aspect to the opposition to poison gas weapons. It was perfectly honorable to use a chemical reaction to launch a projectile into a man's heart or to cause an explosion to blow his legs off, but immoral to use a chemical reaction to destroy his lungs. Although the Hague conventions of 1899 and 1907, at which many nations, including Germany, France, and Britain, agreed to ban the use of asphyxiating gases as weapons of war, were a possible reason for the aversion to poison gas, the novelty of the weapon was more likely the cause of the moral righteousness against its use. "Every new weapon is capable of winning a war," Haber wrote insightfully. "Every war is a war against the soul of the soldier, not against the body. New weapons break his morale because they are something new, something he has not experienced, and therefore, something that he fears. We were used to shell fire. The artillery did not do much harm to morale, but the smell of gas upset everybody."

On April 22, after several aborted attempts because of unfavorable wind conditions, Operation Disinfection, as the experiment had been code-named, was launched. About six thousand man-sized canisters of deadly chlorine gas, 168 tons in total, had been buried in the ground along a stretch of unremarkable countryside near the French town of Ypres. The gas was simultaneously released in a billowing whitish wall that silently seeped toward the

Allied trenches. As it advanced, it rose twenty meters into the air and turned into a great, roiling, greenish-yellow cloud. Many troops retreated, but some disoriented or foolhardy troops charged the cloud. The acrid fumes began to burn their throats. They belatedly stuffed rags into their mouths and wrapped their faces in their shirts or dug holes in the ground and buried their faces in the dirt. The gas burned and then killed as men clutched and clawed at their mouths and eyes. More gas was released over the next several days as the German army sought to advance. The poison cloud was an unexpected success that did indeed clear the trenches temporarily, but the German army was unable to take advantage of the victory because of a lack of troops and munitions. "The generals," writes Jeffrey Allan Johnson, "had thrown away the unique psychological and tactical advantage of attacking an unprotected, unprepared enemy." At the end of the struggle, according to Allied claims, there were 15,000 casualties and nearly 5,000 dead, while the Germans declared that the gas had only affected several hundred German soldiers and killed only a dozen. Both claims were likely exaggerated, the Allies on the high side and the Germans on the low. Although the German troops failed to capitalize on the advantage the chlorine gas had temporarily given them, the ultimate potential of this new deadly weapon was still unknown, and military planners remained optimistic that it might yet provide the means to predictably and consistently clear soldiers from the trenches and end the interminable stagnation of the war. Haber was promoted to captain in the army and placed in charge of Germany's expanding gas warfare and gas defense program.

Allied retaliation with gas weapons came about five months later, as Haber had predicted, but by then both sides had become inured to the use of poison gas. Throughout the war, poison gas weapons became increasingly more sophisticated and advanced, by being launched from exploding shells, for example, rather than allowing for wind drift, and by the development of ever more deadly chemical substances, such as phosgene and mustard gas. Primitive gas masks were developed soon after Ypres, and throughout the war, each new type of poison gas or method of delivery was quickly countered with improved gas masks. By 1917, Haber had fifteen hundred people working under his command and a huge budget, and French and British scientists likewise continued to improve upon the effectiveness of gas weapons and gas defenses. But gas never lived up to its initial optimistic billing. An estimated 125,000 tons of gas had been manufactured and dumped into the trenches and over the battlefields of Europe during the war. Although the various poison gas attacks had created an estimated 1.3 million casualties by the end of the war, this represented less than 6 percent of total war casualties (over 21 million). Poison gas was fatal in 7 percent (approximately 91,000) of those affected by it, while bullets and high explosives killed 25 percent of those injured. Historians generally do not consider poison gas to have been a very effective weapon. Had gas never been used, there is no guarantee that those men wouldn't have been killed by some other more morally acceptable means. Still, 91,000 young men killed in a most horrible manner is no small number, and, rightly or wrongly, Haber was the one held responsible for being the Father of Gas Warfare, for setting the world onto that path, a path that many believed prolonged the war while introducing an additional element of horror. It is worth pointing out that many of the arguments against poison gas warfare were similar to the arguments put forth several centuries earlier against gunpowder weapons, as they were displacing swords, pikes, and longbows and restructuring the nature of combat.

If Haber was ever troubled by his responsibility for the escalation of gas warfare, he never admitted it. He claimed, with perfect logic, that being killed by poison gas was no more abhorrent than being mangled by splinters of steel shrapnel. He never publicly expressed self-doubt or questioned his involvement, indeed pioneering and leading role, in the development of poison gas weapons. Haber's son from his second marriage, L. H. Haber, has written in his thorough study of gas warfare in the First World War, *The Poisonous Cloud,* that his father was "the embodiment of the romantic, quasi-heroic aspect of German chemistry in which national pride commingled with the advancement of pure science and the utilitarian progress of technology. . . . He was a Prussian, with an uncritical acceptance of the State's wisdom, as interpreted by bureaucrats, many of them intellectually his inferiors." Haber did what he was asked to do and left the ethical quandary for others to ponder. He abdicated his moral responsibility. Yet he ultimately paid a high price for his labors. For months Haber's first wife Clara (Immerwahr), herself a highly trained chemist deeply dissatisfied with her role as a housewife in the shadow of her husband's professional triumphs, pleaded with him to forsake his role in what she considered his barbaric and immoral work on poison gas. He persisted, convinced of his overriding duty to devote all his skills and talent to his country during its time of need. Overwhelmed by the force of her husband's character, over the years she had descended from a confident and respected scientific professional before their marriage into melancholy and resentment. Haber was not a supportive husband, nor was he sympathetic to her professional sacrifice and loneliness. He was preoccupied with his own accomplishments, research, and professional reputation. He was an incredibly focused, stubborn, and powerful man. Within days of the gas attack at Ypres, and just as Haber was departing for the Russian Front to supervise more gas installations there, Clara shot herself in the heart with his army revolver, leaving their thirteen-year-old son Hermann to discover her body.

Haber persisted in his work on poison gas throughout the war, pioneering new methods of delivery and new, more powerful toxins. He would not be dissuaded, believing that he was right and others wrong. His patriotic work on poison gas, and above all ammonia synthesis, earned him great respect within Germany: He was showered with national awards, including a knighthood and induction into the Bayer Academy of Sciences in Munich and the Prussian and Göttingen Academies. He was bestowed with several honorary doctorates. He also made a lot of money from his royalties from BASF's ammonia synthesis operations. He was well known, respected, and wealthy, and in 1917 he was married again, to a much younger woman, and started another family, just as his father had done. In his typical brusque and straightforward manner he informed his son Hermann, by then fifteen years old, of the upcoming marriage over dinner on the same night he introduced him to Charlotte Nathan.

By the end of the war, Haber was worn out and despondent over the collapse of Imperial Germany, for which he had devoted years of his life and redirected his career. Even the award of the Nobel Prize was tainted because of the protests of scientific colleagues in Britain, France, and the United States. Protests that, as Haber rightly observed, were a bit hypocritical—poison gas was not the only horror of the war. The Allied, mostly British, blockade of Germany, sometimes called the hunger blockade, denied the German people access to food imports and cruelly starved to death hundreds of thousands during the final years of the war. The deaths—slow, painful, and morale destroying—were for the most part restricted to the civilian population, including babies and young children, who were weakened by malnutrition and made susceptible to diseases. But history is written by the victors and only recently have historians begun evaluating the relative impacts of gas warfare or the starvation blockade on the outcome of the conflict or on the suffering of civilian noncombatants. Logically, of course, Haber was right: It was primarily fear of something new that brought moral condemnation. The destructive capacity of poison gas was no greater than the other forms of modern warfare that were being used to slaughter tens of thousands of young men in the trenches, but the perception at the time was that it was cowardly and inhumane. Haber alone, of the hundreds of scientists from numerous countries who worked on poison gas during the war, became the lightning rod for public condemnation as the Father of Gas Warfare. His own stubbornness likely played a role. "He had done what was in the best interests of his country," writes L. F. Haber in *The Poisonous Cloud*. "The same explanation has been used on many occasions since, and it always appears incomplete . . . governments are unprincipled and excuse their actions on the ground that the national interest calls for them. Individuals aid and abet governments and seek to justify themselves after the event. Regrets or remorse will soften the criticism of history, but Haber was too self-righteous to adopt that line."

Like Alfred Nobel, Haber was a man of sometimes bewildering contradictions. He was known to mock the importance of uniforms and medals, while being slightly embarrassed by his low rank of sergeant and later captain. But he wasn't ashamed to order others to their task when he outranked them. The great irony of Haber's rigid adherence to the belief that he was morally right in devoting himself to his nation's military cause, in spite of the ethical ambiguity of his work, was that he harbored a dislike of military culture. He saw little productive value in war and was particularly disheartened by the disharmony it brought to the international scientific community. But it was the poor planning and colossal inefficiency of the military command that earned his disdain. "The intellect of the army man," he wrote, "trained in the command of troops, lacked the technical imagination to appreciate the changing conduct of war with its technical developments. Without this imagination, preparation followed historic lines. The measure of the needs, and of the methods of fulfilling them, were taken from the experience of the past, in which the technical requirements were different." Yet despite his low opinion of the military and his dislike of virtually all the officers he worked

Three Allied soldiers wear gas masks during the First World War. The photograph shows the ad-hoc and rapidly changing design of new gas defenses from 1915 onward. The German chemist Fritz Haber introduced gas warfare at the Battle of Ypres in April 1915.

with, he energetically devoted his career to the army for the duration of the war. For Haber, duty was paramount. He believed he owed his loyalty to his country, right or wrong.

Haber remained devoted to his country throughout the 1920s, continuing as a director of the prestigious Kaiser Wilhelm Institute in Berlin, which had converted back into a center for peacetime scientific research. One of his great projects during these years was to develop a commercially viable method of extracting gold from seawater. Patriotic to the end, his objective was to help Germany repay her crushing war reparations, which incredibly amounted to about two-thirds of the global supply of gold. Not surprisingly, he failed. The German economy began to collapse and the Weimar Republic disintegrated. The rise of the National Socialist regime and the increasing anti-Semitism was a particularly hurtful development to Haber, the patriot who had devoted years of his life and sacrificed his international reputation for the fatherland. On April 7, 1933, he was ordered to terminate the employment of all his Jewish scientists. Although his conversion to Protestantism, his patriotic war record, and his scientific credentials secured for him a temporary exclusion from the new law, he felt he had no choice but to resign. "My tradition requires of me," he wrote in his letter of resignation, that "in a scientific post in choosing fellow workers I take into account only the professional qualifications and the character of the applicants without asking about their racial disposition. You will hardly expect a man of sixty-five to alter his mode of thought that had guided him in the thirty-nine years of his academic career, and you will appreciate that it is the pride with which he has served his German homeland through all of his life which now dictates the request for retirement."

Broken in spirit and in ill health, Haber considered an invitation to settle in Palestine, but decided to first travel to England, where he was offered an unpaid position in the laboratory of the chemist Sir William J. Pope at Cambridge University. But he found the climate not to his liking, and despite the congenial welcome by many colleagues and old acquaintances (and former enemies), he perceived an aversion to him because of his work on poison gas—now fifteen years in the past. He also struggled with the usual conflicted emotional turmoil of the exile. In Germany he had been famous, respected, honored, and decorated—he was *the* Fritz Haber. In England he was merely a German scientist, slightly tainted with the odium of his wartime history. If he returned to his homeland, he would be just another Jew in Nazi Germany. Earlier in his life, he had shrugged off the insidious influence of anti-Semitism, claiming that it was a privilege because it made Jews work harder. But despite all his efforts to change his ancestry and his triumph over the illogical constraints of racism, in the end he was its victim, driven from his homeland. "In my whole life," he wrote to Albert Einstein in the summer of 1933, "I have never been so Jewish as now." Haber departed Cambridge after only a few months for a respite from the dreary English winter. His health was poor, and he died in Basel, Switzerland, on January 29, 1934. Despite not being very well known outside of academic circles, Haber has had a longlasting and tangible impact on global events that is greater than the work of many of his better-known contemporaries. Not for his work on poison gas, but for the development of the ammonia synthesis process. "There can be no doubt," writes Vaclav Smil, "that Haber's invention of the industrial ammonia synthesis—unchanged today in its fundamentals but greatly improved in its performance—has been one of the cornerstones of modern civilization." It is what won him the Nobel Prize in Chemistry. Haber was chosen by the Nobel Committee not because, as one detractor said, "it prolonged the war," but because synthetic ammonia was the foundation of the global fertilizer industry, which was in turn the foundation of the twentieth century's agricultural revolution, believed at the time to be the solution to world hunger. His discovery was, as stated at the ceremony, "an exceedingly important means of advancing agriculture and the welfare of mankind."

After the war ended in 1918, the technology of the Haber-Bosch process was dispersed to countries around the world. Article 172 of the Treaty of Versailles stated: "Within a period of three months from the coming into force of the present Treaty, the German Government will disclose to the Governments of the principal Allied and Associated Powers the nature and mode of manufacture of all explosives, toxic substances and other like chemical preparations used by them in the war or prepared by them for the purpose of being so used." In 1913 Chile was supplying two-thirds of the global nitrate demand, yet by the 1930s this was reduced to a mere 7 percent, primarily because of ammonia synthesis. The affect was to cause a catastrophic devastation of the Chilean economy and throw tens of thousands out of work. Chile's historic monopoly was broken forever. The Haber-Bosch process had liberated both farmers and warriors from the need to scour the world for natural sources of their most vital raw material.

War and the Green Revolution

"The human race might conceivably blast itself from the face of the earth with nitrogen-bearing explosives, but it will not be starved out for lack of nitrogenous fertilizers."
—Williams Haynes, 1945

Back through the fog of time, sometime in the distant past, around ten to twelve thousand years ago, people somewhere in the Middle East and also in China began cultivating crops for food. Not long afterward an astute man or woman, or perhaps even a child, observed that these early rudimentary crops seemed to grow healthier and larger where animals had defecated. Soon they were collecting the manure of domestic animals and regularly applying it on their crops to improve yields. Animal husbandry and agriculture were linked in a symbiotic cycle where the increased crop yields were fed to the animals, which, in turn, produced manure for fertilization. The link between fertilizer and improved crop yields was understood early in human history, and manure was used for centuries by every culture that relied on agriculture for food, from China to India, Indonesia, the Middle East, Europe, and throughout the Americas.

In addition to animal manure, early fertilizers consisted variously of rotting compost, decaying fish, ground bones, wood ashes, bird and bat guano, chalk, and wool remnants. By the seventeenth century saltpeter was recognized as an excellent chemical fertilizer, easily stored, transported, and applied, with miraculous results, but it never achieved widespread use because of the parallel, and more pressing, demand for its use in explosives and gunpowder. Saltpeter was never available in sufficient abundance for agricultural use until the mid-nineteenth century, when a seemingly limitless organic source of nitrates was discovered in South America. By then, the great benefit of guano and nitrates on crop yields was well known, and these substances were in great demand for use in boosting agricultural production. But after years of intensive cultivation and erosion, soil eventually becomes depleted of nutrients and requires larger quantities of fertilizer to maintain yields. As this happened in the eastern United States and Europe, the application of fertilizers became vital to maintaining a commercially viable agriculture. "Almost every soil type," writes Gilbeart Collings in *Commercial Fertilizers: Their Sources and Use,* "contains limited amounts of one or more of the nutrients necessary for optimum crop growth." Fertilizers return to the soil the nutrients that have been used up by leaching, erosion, and years of growing crops. Nitrogen, phosphorus, and potassium are the three major components of most commercial fertilizers, and the application of these substances, particularly nitrogen, had become ever more necessary with more intensive agriculture and consequent depletion of soils during the late nineteenth century. The limited commercially viable organic sources of these substances, chiefly Peruvian guano and Chilean nitrates, were being depleted at an alarming rate, leading to frightening speculation of a global shortage, which would cause a drastic decline in food production, and that in turn would lead to widespread famine. To compound the problem, nations needed Chilean nitrates to wage a prolonged modern war.

Haber's great triumph was to usher in a new age of bounty, a flourishing of agricultural crops that was believed would avert the impending food crisis and end world hunger. But like the flip side of a coin, synthetic nitrogen compounds have also had an astonishing impact on global population. After the war, when the Haber-Bosch process was adopted throughout the world, synthetic nitrogen began its startling transformation of the planet, not just by unleashing the use of a practically unlimited quantity of explosives in future conflicts, though it certainly did that, but by providing the foundation for the unprecedented growth in global population—more than tripling in under a century from about 1.6 billion people at the turn of the twentieth century to over 6 billion today. It is hard to imagine a more profound and lasting alteration of the culture, political structure, and economy of the world's peoples—the urbanization, industrialization, and terraforming of the landscape to accommodate and feed an additional 4 billion people within three generations. The slow but inexorable taming of previously wild landscapes to accommodate agriculture and the creep of urban sprawl is an ongoing process that in turn creates the capacity to feed more people and hence contributes to further population growth, and the further dependence of our species upon synthetic nitrogen compounds.

More nitrogen compounds are now created synthetically from the air than are available from all the natural microbial processes of the soil. Synthetic nitrogen accounts for about half of the nutrient input into the world's crops and directly feeds billions of people. "For about 40 percent of humanity," writes Vaclav Smil in his thorough and engaging study of nitrogen and global agriculture, *Enriching the Earth,* "it now provides the very means of survival; only half as many people as are alive today could be supplied by prefertilizer agriculture with very basic, overwhelmingly vegetarian, diets; and prefertilizer farming could provide today's average diets to only about 40 percent of the existing population." By the mid-twenty-first-century, the Haber-Bosch process will supply about 60 percent of the world's population with their basic nutrition as the use of synthetic fertilizer is predicted to double in the next half century.

Smil poses the question, what is the most significant scientific discovery of the twentieth century? Many of the common answers are nuclear energy, airplanes, television, computers, or space flight. "Yet none of these inventions has been as fundamentally important as the industrial synthesis of ammonia from its elements," he writes. "Lives of the world's 6 billion people might be actually better without Microsoft Windows and 600 TV channels, and neither nuclear reactors nor space shuttles are critical determinants of human well-being. But the single most important change affecting the world's population . . . would not have been possible without the synthesis of ammonia [which is then converted into nitrogen compounds]." Two-fifths of the world's current population would not be here were it not for the Haber-Bosch process, and the global dependence on synthetic fertilizers will only increase along with global population, which is conservatively estimated to grow to between 9 and 10 billion by the mid-twenty-first-century.

By breaking the symbiotic interdependence of agriculture and animal husbandry, freeing both from a self-regulating nitrogen cycle, synthetic nitrogen compounds have achieved great economies of scale in the production of animal and agricultural products. It has given a greater number of people, and a greater proportion of the world's population, better nutrition than ever before, but with unintended side effects. Now we are faced with the consequences that have arisen from too many nitrates in the environment. After nearly a century, the onetime solution has become another problem.

When nitrogen fertilizers are applied to crops, only a small portion is actually absorbed by the plants. The remainder is washed off the fields and into creeks and rivers that eventually lead to the sea. Currently about 50 percent of the nitrogen fertilizer applied to crops is washed away before being absorbed. Nitrogen not only stimulates plant growth on land but also in the water, and the vast quantities that yearly flood downstream have resulted in large algal blooms on the water's surface, which block out sunlight to other plants and prevent regular photosynthesis. When the algae dies, it sinks to the bottom and decomposes, sucking the oxygen from the water, thereby starving other species, such as crabs and clams and lobsters, of the vital element. Sometimes the algal blooms are toxic and produce what is known as a red tide that kills fish and poisons shellfish. Ironically, the increase in food on land brought about by synthetic nitrogen is slowly causing a decrease in food from the sea. One classic example occurs in the Gulf of Mexico from Louisiana west to Texas where the excess nitrogen runoff in the Mississippi and Atchafalaya Rivers has created a large dead zone, or deoxygenated region, that kills bottom-dwelling species and drives other fish away. Similar negative environmental impacts from excess nitrate runoff have occurred in the lagoon of the Great Barrier Reef off the coast of Australia, which has seen application of nitrogen fertilizers on adjacent farmland increase by a factor of ten in the second half of the twentieth century, and in the Baltic Sea, where the cod fishery collapsed in the 1990s, coinciding with a fourfold increase in nitrogen levels during the same time period.

After nearly a century of unlimited access to synthetic nitrogen compounds and an astounding increase in the domestic crops under cultivation, we now have to contend with the new challenges brought on by that very success.

Only a handful of decades after his world-altering discovery, few people have heard of Fritz Haber or are even aware of how he has profoundly impacted modern civilization. But as with the use of limitless quantities of nitrogen fertilizer in agriculture, the unfettered use of nitrogen-based explosives has also had a mixed record during the century and a half since Nobel invented dynamite. Problems have been solved and problems have been created. The grand benefits in civil engineering and mining can be balanced against the increased destructiveness of wars. It is not a new debate. There has been a great deal of philosophizing throughout our history about the nature and impact of explosives that dates to the earliest use of black powder.

Are explosives a terrible force of evil and destruction? The seventeenth-century alchemist and natural philosopher William Clarke questioned "whether more good or ill has proceeded from this Invention." The English reformer and journalist William Cobbett called gunpowder, along with the banknote, "one of the two most damnable inventions that ever sprang from the minds of men under the influence of the devil." Sebastian Miller in *Cosmographie,* published in Basel in 1584, said of black powder that "The villain that brought upon the earth so injurious a thing does not deserve to have his name remain in the memory of men." Alfred Nobel was known as a merchant of death for his role in pioneering the invention and sale of nitroglycerin and dynamite; he was shunned and threatened with lawsuits and the closure of his businesses. Fritz Haber was criticized for his invention of the process to create synthetic nitrogen compounds because it kept Germany in the First World War. In trying to alleviate hunger and to preserve his country from a humiliating military defeat, he indirectly caused the deaths of millions of young men and the destruction of the European countryside.

On the other hand, explosives have brought much good to humanity, improving the material comfort of billions by enabling travel and trade through roads, railways, and canals and by greatly expanding the capacity for mining and construction. Nobel felt that dynamite and ever more powerful explosives would be a great force of liberation and global peace. The Nobel prize committee awarded Haber the prize in chemistry for his scientific innovation, despite its obvious use in explosives. The nineteenth-century Scottish historian and writer Thomas Carlyle claimed that explosives were "one of the three great elements of modern civilization." And for the opposite reason, to destabilize the current civilization, the anarchist Albert Parsons proclaimed that "in giving dynamite to the downtrodden millions of the globe science has done its best work. . . . It is a genuine boon for the disinherited, while it brings terror and fear to the robbers." Certainly, our modern civilization, complete with its monumental accomplishments and its devastating setbacks, could not exist without the historical and ongoing transformation of the planet through explosives.

Over the centuries explosives have changed how we think. They have expanded our ability to alter the world to suit our needs, from the ways wars are fought to the methods and scope of construction and resource extraction. Yet moral responsibility for harnessing the tremendous and terrible power of gunpowder or dynamite or any other explosive, or even for the unrestricted application of nitrogen fertilizer, has always rested with the individuals who made and continue to make use of them.

High explosives, of course, are neither intrinsically good nor bad, but both, and neither. In the most basic sense, they are nothing other than a tool, a club, or a hammer, but vastly more powerful—an extension of the various and competing motives and desires of the human psyche. The history of the use of explosives through the ages—the moral quandary when they disrupted established social orders, the irony of the dual use of nitrates for war and agriculture, the grasping struggle to control the global nitrate supply, and the lofty dreams of triumphant accomplishment—is nothing other than a history of the conflicted and bewildering duality of the human mind. On the one hand, murderous, frightening, and destructive; on the other hand, optimistic, determined, and wildly inventive.

A Note on Sources and
Further Reading

Because this is intended as a popular rather than scholarly book, I have elected not to include footnotes in the text. Following is a chapter-by-chapter discussion of my main sources for each section, with emphasis on those that would be useful for anyone wishing to learn more on a given topic. A bibliography of key specific and general sources follows. Important contributors are referenced in the bibliography.

I am not a chemist and my explanation of the chemical processes are intended for a general audience. My focus was on the history and impact of the scientific discoveries on world events rather than a detailed explanation of the chemical processes.

CHAPTER ONE: PLAYING WITH FIRE:
A THOUSAND YEARS OF EXPLOSIVES
The history of Guy Fawkes and the Gunpowder Plot is well known and extensively written about. Most libraries should have several books on this topic but the most lucid and comprehensive book on the subject in my opinion is Antonia Fraser's *Faith and Treason: The Story of the Gunpowder Plot*. For copies of original sources see *The Trial of Guy Fawkes and Others*, edited by Donald Carswell.

Roger Bacon is likewise a well-discussed man. Consult his own edited writings with an introductory essay in *The Mirror of Alchemy: Composed by the Thrice-Famous Learned Fryer, Roger Bacon,* edited by Stanton J. Linden. Any general history of science such as George Sarton's *Introduction to the History of Science* will have a section on Bacon. Many histories also have been written on the early history of gunpowder. J. R. Partington's *A History of Greek Fire and Gunpowder,* while being exceedingly dense and scholarly, is loaded with information and a discussion of the early incendiary weapons from around the globe, particularly China and the Middle East. Partington was the source for my quotes on early incendiary and gunpowder weapons. Also of interest is Joseph Needham's *Science and Civilization in China. Ancient Inventions,* by Peter James and Nick Thorpe, includes a great deal of information on early incendiary weapons.

CHAPTER TWO: BLACK POWDER'S SOUL:
THE QUEST FOR THE ELUSIVE SALTPETER
The quest for saltpeter is not a common or popular area of study and most books on gunpowder and explosives will devote at most a paragraph or so to the subject. There are, however, many books dealing with early-seventeenth-century English history and Charles I in particular which discuss the activities of the saltpetermen. Particularly informative is Kevin Sharpe's *The Personal Rule of Charles I,* which provides an excellent discussion of the state-sponsored saltpetermen in England and the critical shortage of saltpeter for gunpowder at the time. For the proclamations in England regarding the preservation of saltpeter beds and the importance of saltpeter and the information on the saltpetermen as agents of the crown see *Stuart Royal Proclamations,* edited by Stuart Larkin and Paul L. Hughes. William Clarke's somewhat amusing seventeenth-century tracts on saltpeter, *The Natural His-*

tory of Nitre, and Thomas Chaloner's *A Shorte Discourse of the Most Rare and Excellent Vertue of Nitre*, can be found on microfilms in the Early English Books Series, a collection of hundreds of old, rare, and obscure English writings that should be available in the collections of most academic libraries.

The saltpeter trade in India has received considerably more attention from historians of Indian history because of its importance to the local economy. Many general histories provide a great overview of the political situation such as John Keay's *The Honourable Company* and Sudipta Sen's *Distant Sovereignty: National Imperialism and the Origins of British India*. Books on the economic activities of the English, French, or Dutch East India companies will usually contain a chapter or at least a reference to the subject, but for a detailed discussion of the saltpeter trade turn to Holden Furber's *Rival Empires of Trade in the Orient, 1600–1800*, Narayan Prasad Singh's *The East India Company's Monopoly Industries in Bihar with Particular Reference to Opium and Saltpeter*, and K. N. Chaudhuri's *The Trading World of Asia and the East India Company, 1660–1760*.

CHAPTER THREE: BLASTING OIL AND THE BLASTING CAP:
ALFRED NOBEL AND THE TERRIBLE POWER OF NITROGLYCERIN
The story of the discovery of nitroglycerin is told in all general histories of explosives such as G. I. Brown's *The Big Bang: A History of Explosives*. See also Cornelius Keleti's collected essays on nitrates in *Nitric Acid and Fertilizer Nitrates*. Greater detail on Nobel's early involvement in explosives research and on the Nobel family in general can be found in Nobel's biographies, particularly Erik Bergengren's *Alfred Nobel: The Man and His Work*, but also Herta E. Pauli's *Alfred Nobel: Dynamite King* and Nicholas Halasz's *Nobel: A Biography*. All of these works are quite old, and it is probably time for a new biography of Nobel, but they all conform in telling the basic

outline of the explosive magnate's life. As Nobel's original documents are largely written in Swedish and are preserved at the Nobel Foundation in Stockholm, I took my quotes primarily from Bergengren, who originally wrote in Swedish and had extensive access to Nobel's personal and business papers. Any good academic library will have in its microfilm or microfiche collection early newspaper accounts such as that from the *San Francisco Chronicle*.

CHAPTER FOUR: CONSTRUCTION AND DESTRUCTION:
DYNAMITE AND THE ENGINEERING REVOLUTION
The story of Nobel's discovery of dynamite is found in all the same sources as the story of nitroglycerin. My two main sources on the history of mining were Gösta E. Sandstrom's *The History of Tunnelling: Underground Workings Through the Ages* and Patrick Beaver's *A History of Tunnels*. My quotes of Agatharchides and Agricola came from Sandstrom. Accounts from newspapers such as the *New York Times* can be found in the microfilm or microfiche collections of any academic library. The complete backlist of *Harper's New Monthly* magazine is likewise available on microfiche.

CHAPTER FIVE: THE GREAT EQUALIZER:
EXPLOSIVES AND SOCIAL CHANGE
The history of the role of gunpowder in changing military structure and social order is a well-researched topic. While not all historians agree on the exact impact gunpowder had in realigning societies in Europe and the Orient, I have confined my discussion to the basic outline of generally accepted outcomes. Although I consulted numerous general histories of warfare I found particularly clear and useful Bernard and Fawn M. Brodie's *From Crossbow to H-Bomb*, Theodore Ropp's *War in the Modern World*, and especially John Keegan's *A History of Warfare*. Two excellent books that discuss social change brought about by explosives are J. D. Bernal's *Science in History* and William H. McNeill's *The Age of Gunpowder*

Empires, 1450–1800. Information on Christian Huygens and Denis Papin comes Joseph Needham's *The Priestly Lecture*.

The basic outline of the Franco-Prussian War is usually related in general histories of warfare such as those mentioned above, while the particular role of Nobel attempting to establish a dynamite factory in France, and a discussion of dynamite's role in speeding the Prussian victory, is discussed in Bergengren's *Alfred Nobel*. Also see *What If? The World's Foremost Military Historians Imagine What Might Have Been*, edited by Robert Cowley, which contains an interesting and informative chapter on the Franco-Prussian War.

CHAPTER SIX: INVENTIONS, PATENTS, AND LAWSUITS: THE GOLDEN AGE OF EXPLOSIVES

Nobel's various biographies were my source for his business dealings in the United States and his overall negative experience there. Bergengren gives more concrete statistics on the wild expansion of Nobel's business empire, the number of factories, their international distribution, etc. These sources also provide basic information on the cordite lawsuit, which is also discussed in most general histories of explosives such as G. I. Brown's *The Big Bang*. Schonbein's discovery of guncotton will also be discussed and outlined in any history of explosives or chemistry and a collection of his own writings on the matter is contained in George Mac-Donald's *Historical Papers on Modern Explosives*.

CHAPTER SEVEN: THE GUANO TRADE: THE TOIL FOR CHILEAN SALTPETER AND THE WAR OF THE PACIFIC

The guano trade is an unusual topic that, not surprisingly, has not commanded a great deal of attention. The best, and perhaps the only, comprehensive account of the whole sordid business is Jimmy Skaggs's *The Great Guano Rush: Entrepreneurs and American Overseas Expansion*. Eyewitness accounts of the guano trade come from George W. Peck's *Melbourne and the Chincha Islands; with sketches of Lima, and a Voyage Round the World* and John Moresby's *New Guinea and Polynesia.* There are other accounts of the cruel and ghastly working conditions, but these were the most revealing and graphic. Information on the impact of the guano trade on Peruvian government finances is discussed in W. M. Mathew's article "A Primitive Export Sector: Guano Production in Mid-Nineteenth-Century Peru" in the *Journal of Latin American Studies,* while information on how the guano was created (ocean currents, bird habitat, etc.), comes from R. E. Coker's article, "Peru's Wealth-Producing Birds," in *National Geographic Magazine.*

Nitrates, what was known as Chilean caliche, have played a greater role in global affairs and have been more commonly studied by historians. M. B. Donald's "History of the Chile Nitrate Industry," in the *Annals of Science,* gives a solid account of the industry throughout the nineteenth century. Also consult Harold Blackmore's *British Nitrates and Chilean Politics, 1886–1896* and Robert Greenhill's article in the *Journal of Latin American Studies* "The Peruvian Government and the Nitrate Trade, 1873–1879." The War of the Pacific is reliably documented in any general history of Chile and Peru, such as William Sater's *Chile and the War of the Pacific.* Statistics on guano and nitrate consumption in global markets come from several sources, the most comprehensive being *The World Fertilizer Economy* by Mirko Lamer and *American Chemical Industry: The World War One Period, 1912–1922* by Williams Haynes.

CHAPTER EIGHT: THE PROFITS OF DYNAMITE:
A GIFT TO SCIENCE AND CIVILIZATION
Alfred Nobel's final years in Italy and Sweden and the disposition of his will is not an obscure topic, but most accounts are superficial. Any of the biographies on Nobel will provide more or less

the same information on his final years and the thoughts that led to his gifting his entire estate to science and civilization. There are dozens of books on the Nobel prizes and the many prize winners. An excellent general and recent source is Burton Feldman's *The Nobel Prize: A History of Genius, Controversy, and Prestige.* The most detailed and authoritative account of administering Nobel's will, *The Life of Alfred Nobel,* by his assistants Henrik Schuck and Ragnar Sohlman, is an intriguing window into the legal difficulties faced by the two men in an era with few precedents for international agreements and without electronic transfers of money. It also details the complicated jurisdictional, legal, and family-related impediments to Nobel's final wishes ever being realized. Bergengren also provides translated quotes from *A Will,* Sohlman's Swedish account of administering Nobel's estate.

CHAPTER NINE: BATTLE OF THE FALKLANDS:
THE STRUGGLE FOR THE GLOBAL NITRATE SUPPLY
The Battle of Coronel and the Battle of the Falklands are well documented naval battles. Two good general sources are Paul G. Halpern's *A Naval History of World War One* and Ronald H. Spector's chapter "The First Battle of the Falklands" in *The Great War: Perspectives on the First World War,* edited by Robert Cowley. John Keegan devotes a chapter to the two battles in *Intelligence In War,* which sheds light on the role of intelligence, particularly the use of wireless transmissions, in the decisions of the commanders.

The critical need to secure access to Chilean nitrates during the First World War is usually only briefly discussed in most histories of the war. For more detailed information consult Grosvenor Clarkson's *Industrial America in the World War: The Strategy Behind the Line, 1917–1918* or Bernard Baruch's *American Industry in the War: A Report of the War Industries Board.* Both these works contain information and statistics on the nitrate situation not just for the United States but for the Allies in general as well as detailed accounts of the political wrangling to acquire nitrates stockpiled by German firms in Chile. See Part Two of M. B. Donald's article "History of the Chile Nitrate Industry" in the *Annals of Science* as well as *The World Fertilizer Economy* by Mirko Lamer and *American Chemical Industry: The World War One Period, 1912–1922* by Williams Haynes for statistics on the increasing annual shipments of Chilean nitrates.

CHAPTER TEN: THE FATHER OF THE WAR:
FRITZ HABER'S WORLD-CHANGING DISCOVERY
Considering his importance in world events Fritz Haber ought to have more written on him. The most recent full biography in English is Morris Goran's *The Story of Fritz Haber*, which is nearly forty years old. Short bios are contained in *Great Chemists*, edited by Eduard Farber, and *Dreams and Delusions: The Drama of German History*, by Fritz Stern, and most recently in Vaclav Smil's *Enriching the Earth: Fritz Haber, Carl Bosch, and the Transformation of World Food Production*. Haber and his work are also featured in numerous books on the Nobel prizes and prizewinners, such as Armin Hermann's *German Nobel Prizewinners: German Contributions in the Fields of Science, Letters, and International Understanding* and Burton Feldman's *The Nobel Prize: A History of Genius, Controversy, and Prestige*.

The impact of Haber's great scientific discovery on the outcome of the First World War has been discussed in several books, including *The Chemical Industry 1900–1930*, written by Haber's son L. F. Haber, and *The Kaiser's Chemists: Science and Modernization in Imperial Germany*, by Jeffrey Allan Johnson. Vaclav Smil's work provides the best technical explanation of how the process actually worked, the science behind it. Anyone searching for a more in-depth discussion of the chemistry should consult Smil's book. Haber's work on poison gas is much commented upon in general histories of the First World War, but the most comprehensive and thorough discussion of gas warfare, and Haber's involvement in it, is *The Poisonous Cloud*, by L. F. Haber.

EPILOGUE: WAR AND THE GREEN REVOLUTION
Any university library will hold several books on fertilizers, such as *Commercial Fertilizers: Their Sources and Use*, by Gilbert Collings, and *Nitric Acid and Fertilizer Nitrates*, edited by Cornelius Keleti. My main source for the quantities of nitrate compounds currently used for agriculture and the problems associated with the exponential increase during the twentieth century was Vaclav Smil's *Enriching the Earth*. Smil also details the increase in global population since Haber's discovery became universally known. The quotations on gunpowder and explosives are taken from J. R. Partington's *A History of Greek Fire and Gunpowder*, from *Bartlett's Familiar Quotations*, and similar sources.

Bibliography

Bacon, Roger. *The Mirror of Alchemy: Composed by the Thrice-Famous and Learned Fryer Roger Bacon,* edited by Stanton J. Linden, Reprint: New York: Garland, 1992. Original printed in London for Richard Olive, 1597.

Baruch, Bernard M. *American Industry in the War: A Report of the War Industries Board.* New York: Prentice-Hall, 1941.

Beaver, Patrick. *A History of Tunnels.* London: Peter Davies, 1922.

Bergengren, Erik, translated by Alan Blair. *Alfred Nobel: The Man and His Work.* London: Thomas Nelson and Sons, 1962.

Berhard, C. G., E. Crawford, and P. Sorbom, eds. *Science, Technology and Society in the Time of Alfred Nobel: Nobel Symposium 52.* New York: Pergamon, 1982.

Bernal, J. D. *Science in History.* Cambridge, Mass.: MIT Press, 1965.

Blackmore, Harold. *British Nitrates and Chilean Politics, 1886–1896.* London: The Athlone Press, University of London, for the Institute of Latin American Studies, 1974.

Brodie, Bernard, and Fawn M. *From Crossbow to H-Bomb.* Bloomington and London: Indiana University Press, 1973.

Brown, G. I. *The Big Bang: A History of Explosives.* Stroud, Gloucestershire: Sutton Publishing, 1998.

Chaloner, Thomas, Sir. *A Shorte Discourse of the Most Rare and Excellent Vertue of Nitre.* London, Imprinted by G. Dewes,

1584, microfilm. Early English Books series, 1641–1700. Ann Arbor: University Microfilms International, 1971.

Charles II, England and Wales, Sovereign. *By the King, A Proclamation Prohibiting the Exportation of Saltpeter.* London: printed by John Bill and Christopher Barker, 1663, microfilm.

Chaudhuri, K. N. *The English East India Company: The Study of an Early Joint-Stock Company, 1600–1640.* London: Frank Cass and Co., 1965.

Chaudhuri, K. N. *The Trading World of Asia and the English East India Company, 1660–1760.* Cambridge: Cambridge University Press, 1978.

Church of England. *A Form of Prayer with Thanksgiving to Be Used Yearly upon the Fifth Day of November.* London: s.n., 1685, microfilm, 1980.

Churchill, Winston S. *A History of the English-speaking Peoples.* New York: Dodd, Mead, 1958.

Churchill, Winston S. *World Crisis*, Part 2; 1916–1918. London: Thornton Butterworth, 1927.

Clarke, William. *The Natural History of Nitre; or, a Philosophical Discourse of the Nature, Generation, Place, and Artificial Extraction of Nitre, with Its Verture and Uses.* England, c. 1670, microfilm. Early English Books Series, 1641–1700. Ann Arbor: University Microfilms International, 1971.

Clarkson, Grosvenor B. *Industrial America in the World War: The Strategy Behind the Line, 1917–1918.* Boston: Houghton Mifflin, 1923.

Coker, R. E. "Peru's Wealth-Producing Birds: Vast Riches in the Guano Deposits of Cormorants, Pelicans, and Petrels Which Nest on Her Barren, Rainless Coast." *National Geographic Magazine*, June 1920.

Collings, Gilbeart H. *Commercial Fertilizers: Their Sources and Use,* 5th ed. New York: McGraw-Hill, 1955.

The Trial of Guy Fawkes and Others (The Gunpowder Plot). Edited by Carswell, Donald. Notable British Trials series. Toronto: Canada Law Book, 1934.

Cowley, Robert, ed. *What If? The World's Foremost Military Historians Imagine What Might Have Been.* New York: G. P. Putnam's Sons, 1999.

Cowley, Robert, ed. *The Great War: Perspectives on the First World War.* New York: Random House, 2003.

Dolan, John E., and Stanley S. Langer, eds. *Explosives in the Service of Man: The Nobel Heritage.* The Royal Society of Chemistry, London, 1996.

Donald, M. B. "History of the Chile Nitrate Industry." *Annals of Science,* vol. 1, no. 1, 1936.

Farber, Eduard, ed. *Great Chemists.* New York: Interscience Publishers, 1961.

Feldman, Burton. *The Nobel Prize: A History of Genius, Controversy, and Prestige.* New York: Arcade Publishers, 2000.

Fraser, Antonia. *Faith and Treason: The Story of the Gunpowder Plot.* New York: Doubleday, 1996.

Frothingham, Thomas G. *The Naval History of the World War: Offensive Operations, 1914–1915,* 3 vols. Cambridge: Cambridge University Press, 1924–1926; reprint Freeport, N.Y.: Books for Libraries, 1971.

Furber, Holden. *Rival Empires of Trade in the Orient, 1600–1800.* Minneapolis: University of Minnesota Press, 1976.

Goran, Morris. *The Story of Fritz Haber.* Norman: University of Oklahoma Press, 1967.

Greenhill, Robert, and Rory M. Miller. "The Peruvian Government and the Nitrate Trade, 1873–1879." *Journal of Latin American Studies,* vol. 5, no. 1, May 1973.

Haber, L. F. *The Chemical Industry, 1900–1930.* Oxford: Clarendon Press, 1971.

Haber, L. F. *The Poisonous Cloud: Chemical Warfare in the First World War.* Oxford: Oxford University Press, 1986.

Halasz, Nicholas. *Nobel: A Biography.* New York: Orion Press, 1959.

Hall, Bret S. *Weapons and Warfare in Renaissance Europe: Gunpowder, Technology, and Tactics.* Baltimore: Johns Hopkins University Press, 1997.

Halpern, Paul G. *A Naval History of World War One.* Annapolis, Md.: Naval Institute Press, 1994.

Hampshire, A. Cecil. *The Blockaders.* London: Kimber, 1980.

Hawkins, Nigel. *The Starvation Blockades.* Barnsley: Leo Cooper, 2002.

Haynes, Alan. *The Gunpowder Plot: Faith in Rebellion.* Stroud, England: Grange Books, 1994.

Haynes, Williams. *American Chemical Industry.* Vol. 2, *The World War I Period.* New York: D. Van Nostrand Company, 1945.

Hermann, Armin. *German Nobel Prizewinners: German Contributions in the Fields of Science, Letters, and International Understanding.* Munich: H. Moos, 1968.

James I, England and Wales, Sovereign. *By the King: a proclamation for the preservation of grounds for making of salt-peter, and to restore such grounds which now are destroyed, and to command assistance to be given to His Majesties salt-peter makers.* London: Bonham Norton and John Bill, Printers to the Kings most Excellent Majesty, 1624. Microfilm.

James, Peter, and Nick Thorpe. *Ancient Inventions.* New York: Ballantine Books, 1994.

Johnson, Jeffrey Allan. *The Kaiser's Chemists: Science and Modernization in Imperial Germany.* Chapel Hill: University of North Carolina Press, 1990.

Keay, John. *The Honourable Company: A History of the English East India Company.* New York: Macmillan, 1994.

Keegan, John. *A History of Warfare.* Toronto: Key Porter Books, 1993.

Keegan, John. *Intelligence in War: Knowledge of the Enemy from Napoleon to Al-Queda.* Toronto: Key Porter Books, 2003.

Keleti, Cornelius, ed. *Nitric Acid and Fertilizer Nitrates.* New York and Basel: Marcel Dekker, 1985.

Lamer, Mirko. *The World Fertilizer Economy.* Stanford: Stanford University Press, 1957.

Larkin, James F., and Paul L. Hughes, editors. *Stuart Royal Proclamations.* Oxford: Clarendon Press, 1973.

Lawson, Philip. *The East India Company: A History.* New York: Longman, 1993.

Leconte, Joseph. *Instructions for the Manufacture of Saltpeter.* Columbia, South Carolina: Charles P. Pelham, State Printer, 1862.

Lloyd George, David. *War Memoirs,* 6 vols. Nicholson and Watson, 1933–36.

MacDonald, George W. *Historical Papers on Modern Explosives.* London and New York: Whittaker and Co., 1912.

Mathew, W. M. "A Primitive Export Sector: Guano Production in Mid-Nineteenth-Century Peru." *Journal of Latin American Studies,* vol. 9, part 1, May 1977.

McNeill, William H. *The Age of Gunpowder Empires, 1450–1800.* Washington: American Historical Association, 1989.

Munroe, C. E. *"The Nitrogen Question from the Military Standpoint." Annual Report of the Board of Regents of the Smithsonian Institution.* Washington, D.C.: Smithsonian Institution, 1910.

Moresby, John. *New Guinea and Polynesia. Discoveries and Surveys in New Guinea and the D'Entrecasteaux Islands: A Cruise in Polynesia and Visits to the Pearl-Shelling Stations in Torres Straits of H.M.S. Basilisk.* London: J. Murray, 1876; reprinted New York: Elibron classics, 2002.

Needham, Joseph. *Science and Civilization in China,* vol. 5. Cambridge: Cambridge University Press, 1976.

Needham, Joseph. *The Priestly Lecture.* London: The Royal Society of Chemistry, 1983.

Padfield, Peter. *The Great Naval Race: The Anglo-German Naval Rivalry, 1900–1914.* London: Hart-Davis, MacGibbon, 1974.

Partington, J. R. *A History of Greek Fire and Gunpowder.* New York: Barnes and Noble, 1960.

Partington, J. R. *A History of Chemistry,* 4 vols. New York: St. Martin's Press, 1961–1964.

Pauli, Herta, E. *Alfred Nobel: Dynamite King, Architect of Peace.* New York: L. B. Fischer, 1942.

Peck, George W. *Melbourne and the Chincha Islands; with Sketches of Lima, and a Voyage Round the World.* New York: Charles Scribner, 1854.

Ramachandran, C. *The East India Company and the South Indian Economy.* Madras: New Era, 1980.

Ray, Indrani. *The French East India Company and the Trade of the Indian Ocean: A Collection of Essays.* New Delhi: Munshiram Manoharlal Publishers, 1999.

Reasons humbly presented to the consideration of the honourable House of Commons, for the passing a bill now depending for the importation of salt-petre occasioned by a printed paper, called The Salt-Petre case. London: s.n., 1693, microfilm. Early English Books series, 1641–1700. Ann Arbor: University Microfilms International, 1971.

Ropp, Theodore. *War in the Modern World.* New York, London: Collier Macmillan, 1962.

Sandström, Gösta E. *The History of Tunneling: Underground Workings Through the Ages.* London: Barrie and Rockliff, 1963.

Sarkar, Jagadish Narayan. "Saltpeter Industry of India." *Indian Historical Quarterly,* vol. 13, 1938.

Sarton, George. *Introduction to the History of Science,* 3 vols. Baltimore: Williams & Wilkins, 1950.

Sater, William F. *Chile and the War of the Pacific.* Lincoln: University of Nebraska Press, 1986.

Schuck, Henrik, and Ragnar Sohlman. *The Life of Alfred Nobel.* London: William Heinemann, 1929.

Sen, Sudipta. *Distant Sovereignty: National Imperialism and the Origins of British India.* New York: Routledge, 2002.

Sharpe, Kevin. *The Personal Rule of Charles I.* New Haven and London: Yale University Press, 1992.

Singh, Narayan Prasad. *The East India Company's Monopoly Industries in Bihar with Particular Reference to Opium and Saltpeter, 1773–1833.* Muzaffarpur, India: Sarvodaya Vangmaya, 1980.

Skaggs, Jimmy M. *The Great Guano Rush: Entrepreneurs and American Overseas Expansion.* New York: St. Martin's Press Griffin, 1994.

Smil, Vaclav. *Enriching the Earth: Fritz Haber, Carl Bosch, and the Transformation of World Food Production.* Cambridge, Mass.: MIT Press, 2001.

Stern, Fritz. *Dreams and Delusions: The Drama of German History.* New York: Knopf, 1987.

Wegener, Wolfgang. *The Naval Strategy of the World War.* Annapolis: Naval Institute Press, 1989. Original published in German in 1929.

Williamson, Hugh Ross. *The Gunpowder Plot.* New York: Faber and Faber, 1951.

Zanetti, J. Enrique. *The Significance of Nitrogen.* New York: Chemical Foundation, 1932.

Index